HOMEGROWN HOLOGRAPHY

HOMEGROWN HOLOGRAPHY

Written and illustrated by **George Dowbenko**

AMPHOTO
American Photographic Book Publishing Co., Inc.
Garden City, New York

Published in Garden City, New York, by American
Photographic Book Publishing Co., Inc. All rights
reserved. No part of this book may be reproduced
in any form without the written consent of the
publisher.

Library of Congress Catalog Card No. 76-16450

ISBN 0-8174-2113-0 (Hardbound)

ISBN 0-8174-2406-7 (Softbound)

Manufactured in the United States of America.

TABLE OF CONTENTS

Appendixes 135

Index

PREFACE

Wading Through Pages of Science Hoodoo

science hoodoo's got the market covered
got more sources of distribution
more insidious more subversive
than the local dealerships of junk

start them young on pre-digested thought
    forms
then from grade school junior senior high
    school college
give them books of programmed garble
guaranteed to stunt the spirit
spawn the one-way channeled mind rot
stop their questions they forget

science hoodoo's whatchacall those
    textbooks
making common knowledge inaccessible
removing it from people's mind grasp
through specialized and secret code words
number symbols without reference
created by the science priestcraft
to confuse what is and to ensure the chaos

these books designed to first discourage
stifle interest with the blandness
halt the mindflow with the nonsense
only words of mental flotsam
incoherent strung together
to distract and shred attention

agents of the science priestcraft:
so-called editors who copy
apelike phrases from other textbooks
change the words around a little
with no understanding of the meaning
and less than interest in the concepts
plagiarism that soothes the ego
deadline's made
the scam is finished

and unconscious tools of priestcraft:
so-called scientists who dabble
try their hand at science writing
with no thought of logic sequence
in their words inept in structure
popular science for the masses
crumbs of babble for the rabble
and the deadliest assumption
that the knowledge fragment needed
to understand the piece of writing
is so obvious well known

sure everybody knows that:
this assumption based on hearsay
out of touch and disconnected
is inherent in their writing
is the cause of widespread dullness
and the ignorance through time

time's up
you know you got to
run the hoodoo out of town

The purpose of *Homegrown Holography* is to expose the reader to a relatively more coherent description of the art and science of holography. Wading through pages of science hoodoo refers to the time spent in reading and rereading, checking and rechecking information about holography and the background subjects of optics, electronics, and physics theory, in general. With the perspective of knowledge gained by learning and teaching others how to make holograms at the School of Holography in San Francisco, California, this mass of accumulated information was correlated and integrated into its present form.

Homegrown Holography naturally begins with a brief description of the apparent qualities and composition of light. No previous technical knowledge on the part of the reader is assumed. Next, the section dealing with the basic concepts of holography is presented in terms simple enough so that only an understanding of the "Light Introduction" is necessary. Finally, the last chapter contains easy-to-follow, step-by-step instructions for making holograms in one's own low-cost holographic studio.

The words, illustrations, and photographs of *Homegrown Holography* are mostly self-defining. The text, together with the pictures, is meant to be a dictionary of the words and concepts of holography and related optics. For the best effect, it should be read in sequence to follow the logical development of the technical terms that are introduced. When a previously unmentioned term is introduced, a definition of that term is provided immediately.

The script of *Homegrown Holography* is composed of two-dimensional drawings, photographs, and words that attempt to describe the process of creating three-dimensional photographic images with laser light. The limitations of describing a hologram to someone who has never seen or experienced one are immediately obvious. To describe a physical event like laser light is just as difficult. All the incredible subtlety and beauty of laser light is reduced to mere words in a description like "pencil-thin red beam of light."

The inadequacy of words to describe the experience of laser light can only be fully appreciated by someone who has seen it and who realizes its uniqueness. So it is in describing a hologram. This difficulty is made clear by the "picture-worth-a-thousand-words" analogy. If a picture is really the equivalent of a thousand words, and a hologram is the equivalent of a thousand pictures, a hologram needs a thousand times a thousand, or a million words, to describe it adequately.

Homegrown Holography was conceived and initially written at the School of Holography in San Francisco, California. It is an absolute pleasure to thank my friends and colleagues for their priceless advice and assistance in this project. Particularly, I would like to thank Lloyd G. Cross, the initiator and general director of the School of Holography, for the time, space, and patience he gave in communicating his knowledge of holography to myself and many others. I would also like to thank my fellow instructors at the School of Holography and co-workers at Multiplex Co. for their invaluable help and interest in the completion of this manuscript, specifically Gary Adams, Pam Brazier, Peter Claudius, Michael Fisher, Michael Kan, Paul Leveque, Lon Moore, David Schmidt, Werner Schultz, and Bob Taunton.

And finally, I would like to express my utmost gratitude and appreciation to all the powers that be which led me and supported me throughout this work. Thy will be done.

GEORGE DOWBENKO

Taos, New Mexico

CHAPTER 1
A LIGHT INTRODUCTION

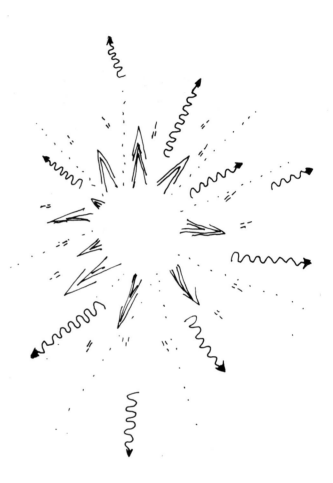

1-1 LIGHT AND COLOR

What is commonly called light is simply a form of physical energy to which the human eye is sensitive. When light energy interacts with matter, the eye receives information about the interaction and transmits this information through the nerves to the human brain. The brain interprets it in the light of past experience and integrates the information in the process called vision.

Vision in humans provides a continuous mapping of the intensity of the light, that is, information about changes in the relative brightness and darkness of the matter. It also provides information about the color of matter as well as three-dimensional information as to the appearance of the structure of the matter itself.

In order to be able to imagine the meeting or interaction of light with matter, some knowledge of the basic structure of matter is necessary. All matter is composed according to a similar pattern of physical energy distribution. This pattern makes the smallest complete energy system of matter grossly comparable to the solar system.

This smallest system, the submicroscopic atomic system, or atom, is structured like the relatively macroscopic solar system. In the center of every atomic system, there is an atomic nucleus just as there is a sun in the center of our solar system. Units of energy, called electrons, revolve through space around the nucleus in constantly changing orbits or circuits. However, like the planets that orbit around the sun, electrons are bound to their circuits so that they maintain the same relative distances between themselves and the nucleus. The nucleus and the innermost electron that whirls around it are comparable to the sun and its innermost planet, Mercury. Like the planets, electrons rotate not only around the atomic nucleus, but also simultaneously spin on their axes.

Whenever several atoms are held together by relatively strong forces, the group of atoms is called a molecule. The

Comparison of an Atomic System and a Solar System

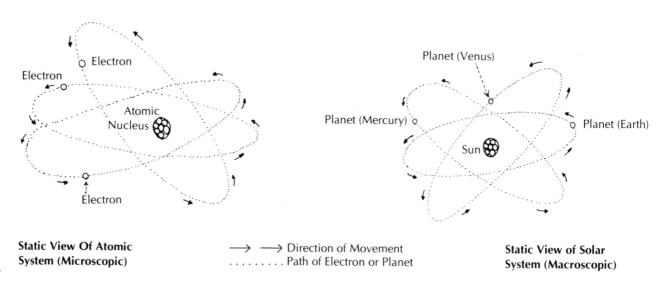

Static View Of Atomic System (Microscopic)

⟶ ⟶ Direction of Movement
.... Path of Electron or Planet

Static View of Solar System (Macroscopic)

(Orbits and relative sizes of attendant electrons and planets are grossly exaggerated.)

light energy interacts with matter, which usually is in this molecular form. The interaction itself involves either absorption or reflection of light energy. When the absorption of light energy takes place, matter assumes the property of color. Color, in turn, is dependent upon the *wavelength* of light that is absorbed by the matter.

Wavelength simply describes the distance that light travels during a certain duration in time. In other words, the wavelength of light is equivalent to the speed at which light travels multiplied by the amount of time involved. Wavelength is usually abbreviated by the Greek letter lambda (λ). This relation can be expressed as:

$$\lambda = v \text{ (velocity)} \times t \text{ (time)} = d \text{ (distance)}$$

When matter absorbs a particular wavelength of light in the presence of a human observer, the human eye receives the impression of a particular color. Each color is measured and classified according to a certain wavelength of light. The unit of measurement is the *nanometer* (abbreviated nm), formerly called the *millimicron* (abbreviated mμ), which is equal to one ten-millionth of a centimeter. Though this unit of length is difficult to imagine, it provides a description of the infinitesimal quality of the energy called light.

The color red refers to wavelengths of light around 650 nanometers. Orange light is measured to be around 600 nm; yellow, 580 nm; green, 500 nm; blue, 480 nm; indigo, 450 nm; and violet, 400 nm. Even though these numbers are virtually meaningless in terms of ordinary everyday description of color, they become valuable in specifying particular hues of color. The light emitted by astronomical objects such as planets and stars, for instance, as well as the light produced by lasers is measured in this way. In this case, the number 6328 is a more accurate description than either of the words red, deep red, or carmine, since the number means a single and unique wavelength of light.

On the molecular level, absorption of light energy takes place when dyes are present in the matter. The dye molecule, excited by the energy of the light, absorbs all the wavelengths of light except the one that yields the actual color of the material. For example, the dye in a red rose absorbs light of all wavelengths except the red. The unabsorbed wavelength is reflected by the rose into the eye of the observer where the sensation of the color red is noted.

On the atomic level, absorption refers to light energy that enters the atomic system and raises the energy level of the entire system. This takes place when the incoming energy is such that an electron is excited and thereby raised to a higher state of energy.

In changing to a higher energy state, the excited electron changes its orbit. Using the solar system analogy, it would be as if the planet Venus changed its orbit to the orbit of the planet Mercury. In other words, a higher energy level means that an electron changes its orbit to one that is relatively closer to the nucleus of the atom. When the atom changes back to the original energy level of the system, it releases energy in the form of the nonabsorbed wavelength of light. This wavelength corresponds to the color perceived by the human eye.

The human eye is sensitive to a continuum of color that is bounded by a deep red at one end and an intense violet at the other. This continuous spectrum of colors is produced by individual and unique wavelengths of light that correspond to the various hues of red, orange, yellow, green, blue, indigo, and violet. The colors simply designate the visible effects of the light energy.

When all the wavelengths of light are emitted or sent out together by a light source, and each wavelength has the same intensity as the others, the result is called white light. Sources of white light include natural sources such as our sun as well as artificial or man-made sources of light such as tungsten-filament light bulbs.

When white light passes through a glass prism, all the colors of light are separated by the prism into the familiar visible spectrum of light. Inside the prism, the white light is simply divided or spread into bands of color that correspond to the various individual wavelengths of light. This spreading effect is called dispersion. Another effect of dispersion in nature is the rainbow. When sunlight, which is white light, passes through droplets of water in the air, which act as prisms, a great curved spectrum is formed in the air. This spectrum, or rainbow, is caused by the breaking up, or dispersing, of sunlight.

In terms of immediate cause and effect, light is the cause and color is the effect of a form of energy in motion. The importance of this energy cannot be overemphasized since its presence is necessary to maintain most forms of life on this planet.

Absorption And Emission Of Light Energy

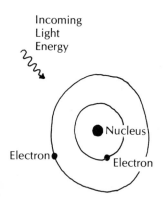

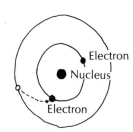

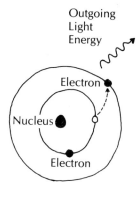

**Atomic System
(Before Absorption Of Light)**

**Atomic System
(During Absorption Of Light)**

**Atomic System
(During Emission Of Light)**

Light enters the atomic system, exciting one of the electrons that orbits around the nucleus. The excited electron is raised to a higher energy level by changing its orbit. Light energy is absorbed by the atomic system, and when the atomic system changes back to its original energy level, energy in the form of light is released or emitted.

The Dispersion Of White Light

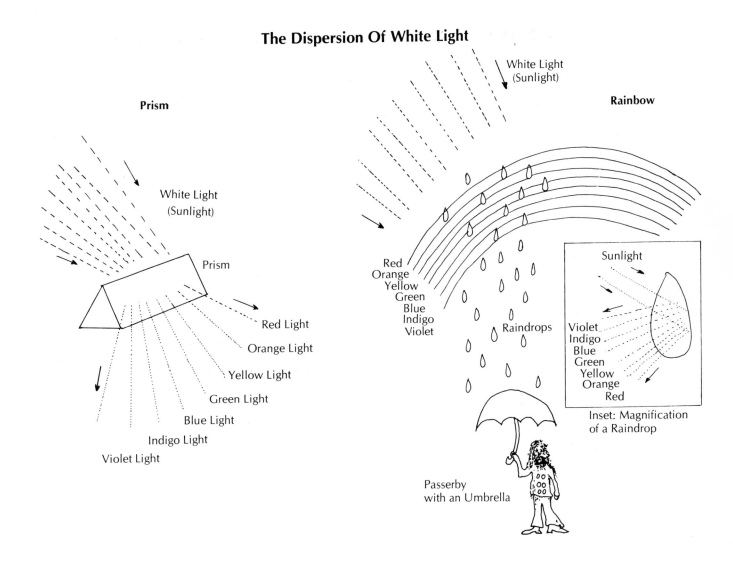

Prism

White Light (Sunlight)

Prism

Red Light

Orange Light

Yellow Light

Green Light

Blue Light

Indigo Light

Violet Light

Rainbow

White Light (Sunlight)

Red
Orange
Yellow
Green
Blue
Indigo
Violet

Raindrops

Sunlight

Violet
Indigo
Blue
Green
Yellow
Orange
Red

Inset: Magnification of a Raindrop

Passerby with an Umbrella

1-2 THE ELECTROMAGNETIC ENERGY SPECTRUM

Light describes only a small portion of the entire spectrum of physical energy known as *electromagnetic energy*. Electromagnetic energy manifests itself simultaneously in its two aspects as electricity and magnetism. These two forms of the same energy in motion, electricity and magnetism, are coexistent through time and space. In other words, they are transmitted together and they also induce one another. The most common example of this is that electric current flow always produces magnetism and magnetism is used to produce electricity.

Electromagnetic energy is found throughout the universe at wavelengths that range from more than 10 megameters in length (10^{10} cm) to less than .001 mÅ long ($^1/_{1000}$ milliangstrom unit, or 10^{-14} cm). The incredibly wide range is due to the various rates of vibration of this energy in motion. Each rate of vibration, whether it's relatively faster or slower, determines the type of electromagnetic energy and its corresponding properties.

The electromagnetic energy spectrum is usually divided to correspond to the different methods known to man of generating as well as detecting this energy. Accordingly, there are different categories of electromagnetic energy and different scales of measurement for each group. The groups of electromagnetic energy in order of decreasing wavelength are as follows: alternating currents; radio or hertzian waves; microwaves; infrared radiation; white light; ultraviolet radiation; X-rays; gamma rays; and secondary cosmic rays, or space rays.

Alternating current (AC) refers to energy with wavelengths ranging from about 10^6 to about 10^9 cm. This energy is the common electrical energy carried by power and telephone lines. It is produced by alternating-current generators, which convert mechanical energy into electrical energy. At its simplest, an AC generator consists of a current-carrying coil or loop of wire that is made to rotate between the poles of a magnet. The mechanical energy that rotates the loop of wire is usually furnished by an engine or an electric motor. As the loop of wire, which carries direct-current (DC) electricity, alternately rotates in one direction then in another in the presence of a magnet, a changing magnetic field is produced. The changing magnetic field induces alternating-current electrical energy. This process is called *electromagnetic induction*.

Electromagnets such as the one described are commonly used to generate alternating current. They are also used to change, or "transform," alternating current in a device called a *transformer*. A transformer consists of two current-carrying coils, one coil having fewer turns than the other. When alternating current passes from one coil to another, the voltage can be raised or lowered to a desired value, depending on the number of turns in each coil: the fewer the turns, the lower the voltage, and vice versa, according to the power requirement.

Electromagnets are also generally used in devices such as the electric bell, the telephone, and the telegraph, as well as in devices that measure electricity, electric

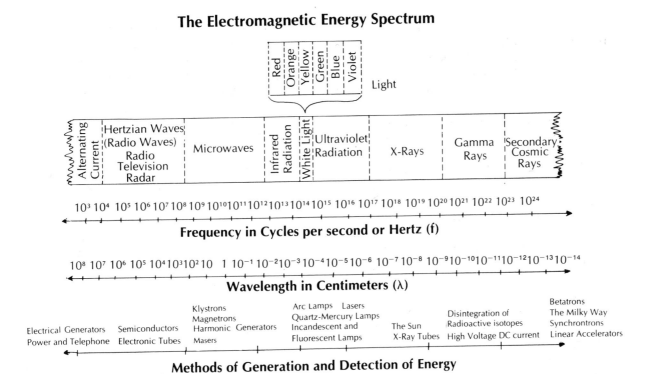

The Electromagnetic Energy Spectrum

meters such as ammeters, voltmeters, and ohmmeters.

Besides wavelength units, alternating current is also measured in *hertz* (Hz), formerly called *cycles per second* (CPS), from about 10 to about 10,000 Hz. This measurement refers to the frequency of an alternating current—the number of complete cycles occurring in each second of time. In the case of an electromagnet, this depends on the speed of rotation of the loop—that is, the number of revolutions per second in a two-pole AC generator.

Hertzian waves can be divided according to radio waves at low, medium, and high frequencies. They also include very high and ultra-high-frequency hertzian waves, abbreviated as VHF and UHF. The lower-frequency radio waves range from about 10^3 to about 10^6 cm wavelengths. Radio waves are generated by electronic tubes and semiconductors. These devices as well as oscillator circuits produce oscillations, or vibrations, in the form of high-frequency alternating current.

In radio broadcasting, this energy in the form of hertzian waves is transmitted by an AC generator that radiates these waves in all directions. The transmitting antenna, which is usually a long wire or a tall, metal tower, emits radio waves that travel through the air until they encounter a receiving antenna. This antenna absorbs energy from the passing waves. The oscillations then are changed into audio information, or sound.

The lower-frequency radio waves are generally used in radio broadcasting while the higher-frequency hertzian waves are used to broadcast the video, or picture, information in television transmissions. The higher frequencies are also used in broadcasting FM radio and radar (an acronym for RAdio Detecting And Ranging). They range from about 10^6 to about 10^8 cm, or 1 to 300 *megahertz*. (The unit megahertz is the same as megacycles per second, or one thousand cycles per second.)

Microwaves are electromagnetic energy at wavelengths from about 1 meter to about 1 millimeter. Microwave-frequency oscillations are produced by electronic devices like the klystron, magnetron, harmonic generators, and also *masers*. (Maser is an acronym for Microwave Amplification by Stimulated Emission of Radiation.)

In a typical ammonia maser, molecules of ammonia gas are excited by microwave radiation. The excited ammonia molecules then collide with nonexcited ammonia molecules and raise them to the same higher energy state. Whenever a collision takes place, microwave radiation is emitted. This interaction is called *stimulated emission,* since the ammonia molecule is in effect stimulated into emitting additional microwave radiation. When enough molecules are stimulated, a chain reaction takes place and the microwave signal is so amplified that it begins generating its own microwave radiation.

Microwaves are generally used in many long-distance communication systems, such as microwave radio relays and others. They have also recently been used in appliances called microwave ovens.

At shorter wavelengths than microwaves but not as short as the wavelengths of light lies the region of *infrared radiation*. The infrared radiation ranges from about 10^{-2} to about 10^{-4} cm, which is between .75 microns and 3000 microns, or 3 millimeters. This type of energy is also called *thermal energy*. It is produced by radiations of heat from the sun as well as quartz-mercury lamps, heat lamps, and other incandescent and fluorescent lamps. Lasers such as the helium-neon laser also produce infrared energy. This type of electromagnetic radiation can be detected by energy-measuring instruments such as the thermopile and also the Golay cell, which is used in astronomy as an infrared detector of celestial objects. Infrared energy, which is not visible to the human eye, literally means below the red portion of the visible spectrum.

Light, or visible electromagnetic energy, lies in a region between approximately 7.2×10^{-5} and 4×10^{-5} cm. Sensitivity to light is dependent only on each individual human's eyes. Beyond these wavelength boundaries, ordinary human vision becomes insensitive to the vibrations of electromagnetic energy.

When the longest of these wavelengths of light, about .00008 cm in length, enters the eye, the sensation of the color red is produced. The shortest wavelength of light, about .00004 cm, produces the sensation of violet, and the wavelengths in between account for the intermediate colors of orange, yellow, green, blue, and indigo.

Light is generated either by artificial

light sources or natural sources of light. Artificial light sources include incandescent and fluorescent lamps, arc lamps, and lasers. Natural sources include fire and our sun.

An example of the incandescent lamp is the ordinary light bulb. The light bulb consists of a glass-enclosed tungsten filament that is heated by the flow of alternating current until light is emitted. Fluorescent lamps also use an electric discharge to produce light. Electricity is used to activate a low-pressure mercury arc lamp to emit ultraviolet (UV) energy. The UV energy in turn excites the fluorescent powders that coat the inside of a glass tube and light is emitted in the process.

Lasers, unlike other sources of light, do not produce white light, which is a mixture of all wavelengths. Instead, they emit light only at single and specific wavelengths, which are particular colors. A helium-neon laser, for instance, only emits red light; argon lasers produce individual lines of green and blue.

In general, light itself is usually channeled and controlled by lenses and mirrors. It is utilized as a source of illumination in both photography and holography. It accounts for not only human vision, but also the vision of animals as well as animal and plant seasonal movements and all types of photosynthesis. All in all, light is indescribably important since most life forms are dependent on it for their existence.

Ultraviolet radiation is electromagnetic energy of wavelengths shorter than light. These range from about 10 to about 400 angstrom units. (An angstrom unit is equal to 10^{-8} cm.) Ultraviolet rays are also called the chemical rays of sunlight since this energy is absorbed and emitted in many chemical reactions. It is also absorbed and emitted in interatomic energy transitions when an atomic system changes its energy level.

Ultraviolet radiation is generated by the sun and also by fluorescent lamps, arc lamps, and lasers. Although humans are ordinarily insensitive to ultraviolet radiation, there is increasing evidence that the pineal gland, the so-called third eye, is an atrophied if not dormant sense organ for UV energy.

X-rays are electromagnetic radiations at wavelengths even shorter than UV rays. Correspondingly, X-rays have higher frequencies, that is, more vibrations per second. This region of radiation ranges from about .1 Å to about 400 Å, or 10^{-6} to about 10^{-9} cm.

Solar X-rays are emitted by the sun, and artificially generated X-rays are emitted by X-ray tubes. An X-ray tube consists of a glass tube that contains a coil of tungsten wire. The tungsten wire is heated by a high-voltage electric current so that electrons are emitted. These electrons are then suddenly stopped by another piece of tungsten in their path inside the X-ray tube. When the electrons are stopped, some of the energy changes into heat and the rest becomes X-rays.

Because of their short wavelength, X-rays are able to penetrate materials through which light will not pass. This property of penetrating opaque materials makes them useful in medicine, dentistry, and crystallography. By using X-ray photography, the condition of bones, internal organs, and teeth can be ascertained. In X-ray crystallography, X-rays are used to determine the geography, or placement, of atoms within the crystal itself.

Gamma rays refer to electromagnetic radiation ranging from about $1/10$ angstrom to 1 milliangstrom ($1/1000$ angstrom). In standard units of measurement, the wavelengths are unimaginably small, from 10^{-9} to about 10^{-11} cm. Gamma rays are usually abbreviated by the Greek letter gamma, as γ rays.

The emission of gamma rays occurs with the more or less spontaneous dissociation of atomic matter. The disintegration of radium and radioactive isotopes produces this form of energy. Gamma rays are also produced by X-ray tubes and high-voltage DC generators. Because of their extremely short wavelengths, gamma rays are also able to penetrate solid matter, such as several inches of solid lead.

Secondary cosmic rays, or space rays, are the shortest known electromagnetic emanations, with wavelengths that are less than 1 milliangstrom. They exist at such a high rate of vibration and are so short-lived that they can be detected only by the tracks they leave. These tracks, which last for only infinitesimal durations of time, are the only physical proofs of their existence. This type of radiation is produced during the creation of matter in its stable atomic form and also during the disruption of the atomic structure. The process is known as emission by elementary particles.

Secondary cosmic rays refer to gamma rays that are produced by cosmic rays. A natural source of these emissions is the Milky Way. They can also be produced artificially by linear accelerators, betatrons, and synchrotrons, which accelerate atomic particles to extremely high speeds until this radiation is emitted. The energy of these short wavelength emissions is measured in millions of electron volts.

The description of the various known forms of electromagnetic energy shows that it manifests itself in a spectrum that includes not only the super macroscopic, such as radio waves several miles in length, but also the submicroscopic, gamma rays several billionths of a centimeter in length. The difference in the scale of the entire electromagnetic spectrum is completely awe-inspiring. What's even more amazing is that human beings are directly sensitive to only one octave of the whole spectrum—light, or visible electromagnetic energy. All the other forms of this energy were detected indirectly or through man-made instruments.

1-3 THE WAVE DESCRIPTION OF ELECTROMAGNETIC ENERGY

Since only the effects of light and not light itself are directly observable, light can be described only in terms of its apparent behavior. This behavior, the apparent wave motion of light, is the basis for the wave description of electromagnetic energy.

The wave description of electromagnetic energy is a model that attempts to explain light in terms of its motion. In this sense, a model refers to a set of ideas and concepts that describe a particular physical event such as light. These concepts usually evolve in the course of time, depending upon observations and experiments dealing with the phenomenon. Also necessary is a general belief in the accuracy and relative truthfulness of the resulting description. When previously unknown information is revealed, which does not conform to the model, the model, together with its foundation, the theory, is either discarded or in some other way displaced. The so-called wave description of light is also in a constant state of revision. As such, it remains, in mathematical terms, only an approximation of the reality that is light.

When electromagnetic energy passes through space and time, it appears to travel in the form of a wave. The form of the energy in motion, which is the wave, should not be confused with its substance, the energy itself. In other words, a simple description of the path of energy as it travels through space and time is a wave. The appearance of this wavelike motion is due to the resistance of various forces to the passage of electromagnetic energy. If it weren't for this resistance, the energy would travel along the path of a straight line.

Though it is not electromagnetic, sound is another form of physical energy that travels along a wave path. Sound waves are produced by the vibrations of such matter as vocal cords in the case of speech and song. The vocal cords vibrate, causing the air molecules in turn to vibrate a certain number of times per second. In this way, sound waves are carried through space. This frequency of vibration is commonly called the audio, or audible, frequency.

Another example of the transfer of energy is the movement of water waves. Because of their resemblance in wave mo-

tion, moving water, which is actually a gross physical event, and the relatively subtler passage of light are often compared to one another. Even though there is a kind of shallow analogy between water waves and light waves, any such comparison can be regarded only as unconscious poetic license. Except in the most general similarity of each apparent wave motion, any comparison of the two is bound to remain superficial.

For all practical purposes, all electromagnetic energy is transmitted throughout the universe at the constant velocity of about 300,000 kilometers per second, or roughly 186,000 miles per second. This constant velocity is commonly called the speed of light and is abbreviated by the letter "c."

At its simplest, the motion of light through space and time can be described as the motion of two sine waves at right angles to each other and to the direction of propagation, or movement. The motion of one sine wave represents the electrical force or field and the motion of the other represents the magnetic force or field. In this sense, an electrical or magnetic field simply designates a space in which electrical or magnetic force is active.

The two energy fields, electrical and magnetic, are not only coexistent and co-inducing, but also propagate together through space and time. According to the wave description of electromagnetic energy, this combined energy field in motion is called an *electromagnetic wave*. At visible wavelengths, it is called a light wave.

An electromagnetic wave can be graphically represented by two sine curves at right angles to each other. The area under one sine curve represents the elec-

An Electro-Magnetic Wave
(According to the Wave Description of Light)

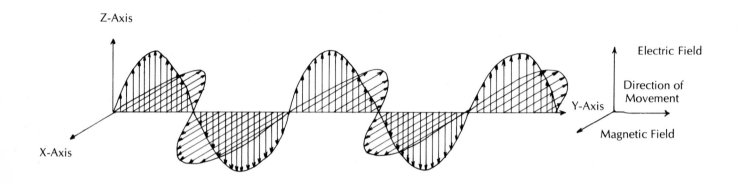

The electric field is represented by the vertically striped areas under the sine curve; the magnetic field is represented by the diagonally striped areas. These two curves represent an electro-magnetic wave at one instant in time. The electric and magnetic fields are at right angles to each other and to the direction of movement, which is the Y-axis.

tric field—space containing electric force in constant motion. The area under the other sine curve likewise represents the magnetic field.

Since the interaction of light with a light-sensitive recording material, such as photographic or holographic film, depends only on the electric field, the magnetic component of an electromagnetic wave can be disregarded. This in turn simplifies the picture so that the electric field in wave form can be described by a single sine wave. The sine wave represents what is called the *electric potential*.

The electric potential of electro-magnetic waves is usually measured in terms of alternating-current (AC) voltage. Voltage simply designates electromotive force or electricity that moves in a current. Voltage is also a relation of the difference in electric potential.

At any instant of time, the measured voltage corresponds to a positive value (+), or a zero value, which is no voltage, or a negative value (−).

By plotting the instantaneous measured voltages of an alternating current, a sine curve is produced. The sine curve rep-

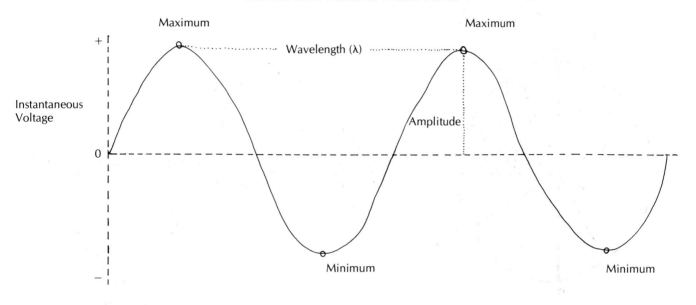

The Electric Field in Wave Form

The electric field is represented by a sine curve. The sine curve represents the electric potential, which is measured in terms of AC voltage. The instantaneous voltage corresponds to a positive (+), a negative (−), or a zero (0) value at any given time.

resents an alternating current whose intensity is constantly changing. The intensity changes from a zero value to a maximum positive value, called a *maximum*, back to a zero value, then to a maximum negative value, called a *minimum*, back to zero, and so on.

The electric field in motion can be visualized as the vibration, or oscillation, of an electric charge at many trillion times per second. The sine curve itself, then, is simply a static, two-dimensional picture of electrical energy vibrating at an incredibly high frequency.

Graphically, the sine curve is obtained by plotting the sine of an angle against its measurement in degrees of rotation (°). This angle is produced by a vector that rotates counterclockwise through 360°. The vector is called a *voltage vector* because it represents the exact instantaneous value of voltage. When the vertical component of this rotating vector is plotted against each angular position, the resulting graph is a sine curve. The rotating voltage vector, then, is simply a graphic representation of a current-carrying coil that also rotates counterclockwise between two

The Development Of A Sine Curve From A Rotating Vector

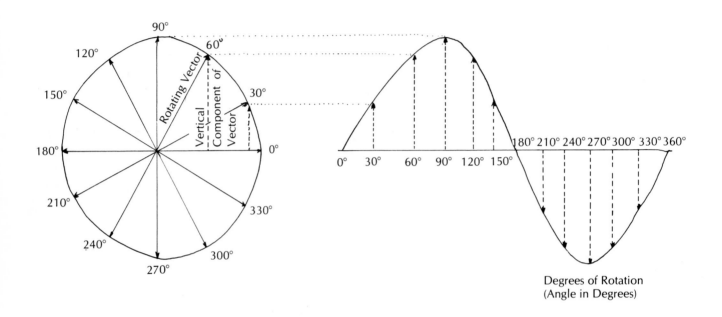

The vertical component of a rotating vector is plotted for each angular position, resulting in a sine curve.

magnetic poles. This rotation generates AC electricity.

A single complete rotation of either a coil or a vector from 0° to 360° is called a *cycle*. One complete cycle also refers to a sine curve that passes from a zero value to a maximum positive value, back to zero, to a maximum negative value, and back to zero again. A cycle, however, does not necessarily begin and end at the zero value. It can be measured from any point along the sine curve to a point that corresponds to a rotation through 360°.

The number of cycles per second is called the *frequency*. Frequency is abbreviated by the Greek letter nu (ν) and is generally measured in hertz units. The frequency of light, for example, ranges from about 4300 to 10,000 trillion cycles per second. In terms of wavelength, this range corresponds to about 7700 to 3900 angstrom units (or ten billionths of a meter).

The frequency and wavelength of light are related to each other by the expression:

$$\underset{\text{(frequency)}}{\nu} \times \underset{\text{(wavelength)}}{\lambda} = \underset{\text{(speed of light)}}{c}$$

which is read: *Frequency multiplied by the*

Phase: In And Out

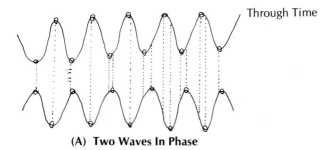

(A) Two Waves In Phase

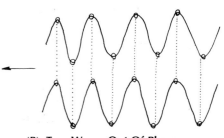

(B) Two Waves Out Of Phase

When the maximum points and minimum points of any two waves coincide (A), the two waves are said to be "in phase." When the maxima and minima of two waves do not coincide (B), the two waves are "out of phase."

wavelength is equivalent to the speed of light. Since the speed of light remains a universal constant, this relationship shows that the frequency of light is directly proportional to its wavelength. In other words, as the frequency of light increases, the wavelength decreases, and vice versa.

Just as the wavelength of light corresponds to its color, so the intensity, or brightness, of light depends upon its *amplitude.* Amplitude refers to a maximum value of the voltage. A large amplitude, then, indicates a high intensity, or brightness, of the light source.

When the maximum points on the sine curve (the maxima) and the minimum points on the sine curve (the minima) of any two electromagnetic waves coincide, the two waves are said to be *in phase.* When one wave reaches its peak, or maximum value, at a different time than the other wave, the two waves are said to be *out of phase.*

Because of the many visible effects of light, which can be explained in terms of light waves, the wave description of electromagnetic energy is most commonly used to account for the behavior of light.

1-4 QUANTUM DESCRIPTION OF ELECTROMAGNETIC ENERGY

Even though the wave model of electromagnetic energy describes the apparent motion of light through space and time, it doesn't account for the actual composition of light in terms of its energy. Neither does it describe the energy transfers that take place when light interacts with matter. In other words, it doesn't explain how light is absorbed or emitted from matter. The quantum description of electromagnetic energy, however, alleviates exactly these shortcomings.

Although light seems to travel in the form of a wave, its wavelike path is just an appearance. Since light is simply a succession of very rapidly moving particles of energy, its wavelike path is only the reaction of certain forces in space to its passage. If it weren't for the influence of gravity and other forces, the definite energy particles that make up light would proceed in straight and direct lines through space. However, because of the resistance of space to the passage of high-velocity beams of energy such as light, it appears to travel in wave form. This appearance is most obvious in optical effects like diffraction and interference.

The distinct particles of energy of which light is composed are usually called *photons*. Photons are energy units of light just as electrons are energy units of matter. When matter absorbs or emits light, the energy of a photon is described as a *quantum* of light. Usually, photons are used to describe the actual physical units of light while quanta are used to describe light in terms of energy exchanges during the interaction of light and matter.

The energy content of one quantum of light is called the photon energy. Photon energy is measured either in joules, which are also called watt-seconds, or electron volts. An electron volt designates a unit of energy that an electron acquires by passing through a potential difference of one volt.

One quantum of electromagnetic energy, that is, the energy of a photon, is related to the frequency of the radiation by the relation:

$$E = h\nu$$

when E is equal to the photon energy, h is equal to a universal constant known as Planck's constant, and ν is equal to the frequency of the radiation. Planck's constant is equal to 6.6×10^{-34} joule-second and is considered to be a fundamental constant value of nature.

When light interacts with matter, the quantity of energy absorbed or emitted by the atomic system is always one quantum or some whole-number multiple of this basic unit of energy. An atomic system that absorbs or emits light energy is said to either raise or lower its quantum level. The field of study that deals with energy exchange in terms of quantum levels is called *quantum mechanics*.

According to the relation $E = h\nu$, the energy of a photon is directly proportional to the frequency of the electromagnetic radiation. In other words, a photon of a higher frequency has a higher basic energy content than a photon of a lower frequency.

In order to demonstrate the energy transactions that take place when light comes into contact with matter, it is necessary to describe the energy content of matter. The energy inherent in matter, that is, in any atomic system, is related to the mass of the matter and the speed of light in Einstein's well-known equation:

$$E = mc^2$$

when m is the abbreviation for mass and c is the abbreviation for the speed of light. This relationship indicates that the energy in any atomic system is simply equivalent to its mass when it is accelerated to the square of the speed of light.

When the two energy relations of matter and light ($E = mc^2$ and $E = h\nu$) are combined, the equivalence of matter and light in terms of energy becomes obvious:

$$mc^2 = h\nu$$

Because of this energy relationship, the mass of any matter can be described in terms of its frequency of vibration. Also, the frequency of a photon can be described in terms of its mass. What this implies is that the physical energy, which is matter, is simply the same energy as light except at a lower frequency of vibration.

The most evident difference between light and matter, however, is that the velocity at which light travels is always greater than the velocity of matter. This is true in all cases except when matter in the form of elementary particles is accelerated to a velocity that approximates the speed of light. When this occurs, mass is abruptly reduced as are the influences of gravity, temperature, and energy currents, which normally act upon the mass itself.

The mass of light, or electromagnetic energy in transit, however, is relatively negligible so that gravity and other forces have little or no influence. Matter in the form of an atomic system, though, is held together by gravity and by other interatomic forces. If it were not for these delicately balanced forces, an atomic nucleus surrounded by orbiting electrons would be unable to exist in this form.

The energy and stability of any atomic system depend not only on the revolutionary rates of all its components—the electrons, protons, neutrons, and other energy units—but also on the number and size of the electrons as well as their distance from the nucleus.

The quantum description of electromagnetic energy reinforces the assertion that the energy inherent in matter is essentially the same as the energy of light. Since both electronic energy and photon energy can be measured in terms of quanta, the interaction of light with an atomic system as well as interatomic energy transactions can be quite accurately described.

1-5 THE USEFULNESS OF WAVES AND PHOTONS

Despite the mathematical elegance of the theory and the relative ease with which the passage of light may be visualized according to the wave description of electromagnetic energy, the description of light as a wave in motion is after all a gross simplification of the actual physical event. Graphically, the sine curve, which represents a sine wave, which in turn represents the oscillations of electromagnetic energy, is actually as much an allegory as it is a map of a three-dimensional event. The situation is further complicated by the fact that the oscillations of light energy take place at frequencies that not only are impossible to observe directly, but are difficult to imagine as well. These are the frequencies of several trillions of vibrations per second.

Likewise, the theory that is the basis for the wave description of light involves assumptions and ideas which are not directly derived from experience. In other words, human knowledge of light depends upon the observation of the effects of light, such as diffraction and interference patterns, rather than on the observation of light itself.

The wave description of light, however, is quite adequate in describing and explaining the optical effects of light, namely the appearance of diffraction and interference patterns. These phenomena are due to the interaction of light waves with other waves of light.

The wave model of light, then, is able to account for the passage of light through space and time as well as describe the interaction of light.

Energy transactions of light with matter and energy transitions within the atom itself can both be represented in terms of the quantum description of light. This quantization of energy, that is, the division of energy into individual units called quanta, accounts for the emission as well as the absorption of light within an atomic, or molecular, system. In other words, there is a definite procedure that is enacted each and every time electromagnetic energy is absorbed or emitted by matter. The description of this procedure—the transfer of energy in quantum units—is the basis for the quantum description of electromagnetic energy.

Just as light and matter are related to each other by the equivalence of their energies, so are the wave and quantum aspects of the description of light related to one another by their equivalence to Planck's constant, h. In other words, the description of light in terms of wave motion and the description of light as quantum units of energy are not separate, unrelated, or conflicting theories, but are descriptions of the same phenomenon from different perspectives.

Even though they are different perspectives on what exactly is light, the wave and quantum qualities of light are inextricably linked. Photon energy and the momentum of a photon are both attributes of a photon in the quantum description of electromagnetic energy. Frequency and wavelength of radiation are both attributes of the wave description of electromagnetic energy. All these attributes are in one way or another related to Planck's constant and therefore are related to one another. However, when the photon attributes of light have large numerical values, the wave attributes are correspondingly small, and vice versa.

Both wave and photon properties of light can never be simultaneously measured by experiment even though their values can be deduced by the knowledge of one or the other. In other words, both the momentum and the exact position of a photon in space cannot be simultaneously determined by experiment. If an experiment is designed to measure one of these properties of light, the other value will become uncertain, and vice versa. This finding is aptly called the *uncertainty princi-*

ple. Since both of these properties cannot be simultaneously determined, this law is also called the *principle of indeterminacy.*

In general, the wave description of electromagnetic energy is useful in characterizing the lower-frequency portion of the spectrum ranging from light to the radio-wave region. Energy at the higher frequencies—light through the X-ray and gamma ray regions—behaves more in accordance with the quantum description of energy.

Light, however, is the only radiation in which both the wave attributes and the photon description are equally valid. The appearance of interference and diffraction patterns testifies to the relative accuracy of the wave description of light. The measurement of light as photon energy and the electronic energy transitions of light and matter support the quantum description of light.

The wave and photon descriptions of light are simply complementary ways of describing the same phenomenon. This is called the *principle of complementarity* since both descriptions are necessary to obtain a relatively complete description of light.

Interestingly enough, even though the complementarity principle requires both descriptions of light for a complete overview of the event, the uncertainty principle prevents the simultaneous knowledge of both descriptions. This means that no matter what kind of experiment is devised, the observer of the event of light cannot disassociate himself from the event. The observer becomes part of what is being observed since the event of light is changed and altered simply by the observer's presence. In other words, the uncertainties of the experiment are impossible to evaluate without changing and thereby spoiling the original conditions under which the experiment was conducted. Because of this, the existence of the two complementary descriptions of light, though really useful for most occasions, is ultimately based on a simple paradox.

It should be kept in mind that all energy in the form of light, heat, electricity, magnetism, and other forms, including matter, are one and the same phenomenon. However, because of the constantly changing and never-ending transformation of energy and matter, physical energy is able to manifest itself in so many different forms. These forms of energy range from everyday and mundane objects of matter to energies which are invisible to the human eye. Matter is, after all, simply a humanly perceptible energy pattern that appears to be more or less stable during the moment of perception. Light, however, is just a taste more evanescent.

CHAPTER 2
THE HOLOGRAPHY RAP

2-1 PHOTOGRAPHY AND HOLOGRAPHY COMPARED

Photography and holography are both methods of recording the various intensities of light from an illuminated subject. Intensity of light refers to the relative brightness and darkness of the subject. When the subject is lit up, the recording takes place as light from the subject reflects onto a light-sensitive material. A light-sensitive, or *photosensitive*, material is a substance that reacts chemically to the presence of light. This chemical reaction between light and a photosensitive material produces a permanent record of the relative brightness and darkness of the subject. The permanent record of this transaction is called a *photograph*, in the case of photography, and a *hologram*, in the case of holography.

The light source used for the illumination of the subject is different in each technique. In holography, relatively coherent light, such as the light from a mercury arc lamp or preferably a laser, must be used. In photography, however, natural sunlight or incandescent lighting is quite adequate.

The photosensitive material used in both techniques is usually the same, most commonly a form of silver-halide emulsion. Silver-halide emulsion consists of microscopic crystals of silver bromide or silver chloride suspended in gelatin. When this emulsion is coated onto glass plates or film, the result is called either photographic or holographic plates or film.

Making a photograph simply consists of illuminating the subject with sunlight or artificial light so that the light from the subject is reflected onto a photographic plate or film. In order to expose the photographic film for only a particular duration of time, the light from the subject must pass through the shutter that opens and closes the aperture. The aperture, which is also called a stop, is an opening that effectively stops all extraneous light except for the subject light. The aperture is a pinhole in its simplest form. After the aperture, the

Making A Photograph With A Photographic Camera

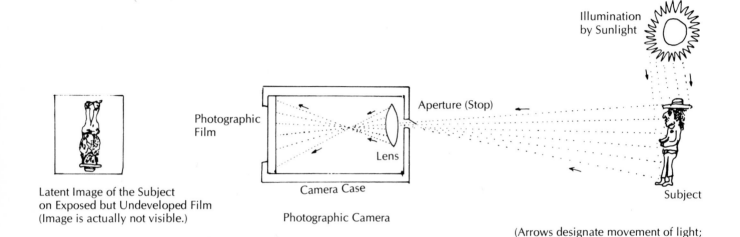

Latent Image of the Subject
on Exposed but Undeveloped Film
(Image is actually not visible.)

Photographic Film

Camera Case

Photographic Camera

Aperture (Stop)

Lens

Illumination by Sunlight

Subject

(Arrows designate movement of light;
dots designate path of light.)

Front View of Photographic Film

Side View of Photographic Camera

The subject is illuminated by light from the sun. This light reflects from the subject, enters the photographic camera, and exposes the film, forming a latent image of the subject.

subject light passes through a lens. The purpose of the lens is to focus the light from the subject onto the photographic film. In focusing the subject light, the lens also reduces the three-dimensional information from a real object to a flat, two-dimensional photographic image. The encased shutter, aperture, lens, and photographic film are collectively called a photographic camera.

There is no visible effect on the photographic film after exposure to the subject light. However, there is an invisible latent image that develops into a visible image after wet processing. Wet processing

refers to the solutions of developer, stop, and fix that act on the exposed photographic film.

After processing and printing the image on photographic paper, the photograph is ready for viewing. To view a photographic print, sunlight or artificial light illuminates the photograph, which reflects the image into the eye of the beholder.

A photograph basically consists of light and dark areas that the eye receives and the initiated mind integrates into a two-dimensional representation of a three-dimensional subject. However, there

Processing The Latent Image And Printing A Photograph From The Negative

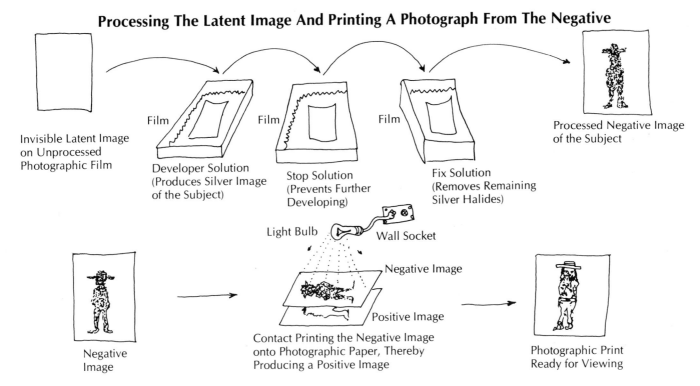

Invisible Latent Image
on Unprocessed
Photographic Film

Film

Developer Solution
(Produces Silver Image
of the Subject)

Film

Stop Solution
(Prevents Further
Developing)

Film

Fix Solution
(Removes Remaining
Silver Halides)

Processed Negative Image
of the Subject

Light Bulb   Wall Socket

Negative Image

Positive Image

Negative
Image

Contact Printing the Negative Image
onto Photographic Paper, Thereby
Producing a Positive Image

Photographic Print
Ready for Viewing

is absolutely no physical resemblance between a two-dimensional photograph and the subject it attempts to depict. The ability to discern information from the seemingly random array of black-and-white spots that make a photograph must be learned. This learned ability is another technique of organizing and integrating certain impressions of the senses. In other words, the method of seeing that is called "reading a photograph" represents material reality only as a map represents actual physical terrain.

The areas of light and darkness in a photograph correspond to the different intensities of light reflected from the subject. They are actually areas of white, shades of gray, and black. Because the lens in the camera reduces the subject light to a flat, two-dimensional image, there is exactly a 1:1 correspondence between each point of light and darkness on the illuminated subject and each point of light and dark on the photographic image. In other words, each point of light on the subject corresponds to one, and exactly one, white or gray area on the photographic print. Likewise, darkness from the subject translates as a black area in the photographic image.

One of the limitations of the photograph is that the three-dimensional qualities of the subject are reduced to only two dimensions: length and width. The dimension of depth is eliminated. Another limitation is that a photograph is unable to record *parallax*. Parallax refers to the apparent shift in perspective that occurs when the subject is observed from one position

Viewing A Photographic Print

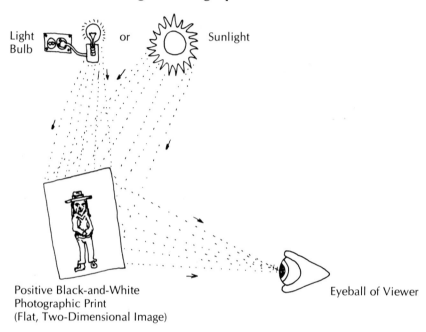

Sunlight or artificial light illuminates the photograph, which reflects light and dark areas
(the photographic image) into the eye of the beholder.

and then from another. This change in perspective takes place when the eye changes its position with respect to the subject. For instance, when the subject is seen with one eye and then with the other, the apparent shift of the subject is called parallax. Because of the two-eyed vision of humans, parallax and the appearance of a three-dimensional world are standard sense impressions that are lost in the technique of photography.

In contrast to ordinary photography, no image-forming lens is used in *holography.* For this reason, holography has been called *lensless photography using laser light.* (*Laser* is an acronym for Light Amplification by Stimulated Emission of Radiation.) Instead of the 1:1 correspondence between subject and photographic image, each point of light from the illuminated subject is recorded on the entire holographic plate. In other words, each piece of the holographic plate receives information from the entire illuminated subject. This is made possible by using coherent light to illuminate the subject. When coherent light in the form of laser light is used, an *interference pattern* is recorded on

Making A Hologram With A Holographic Camera

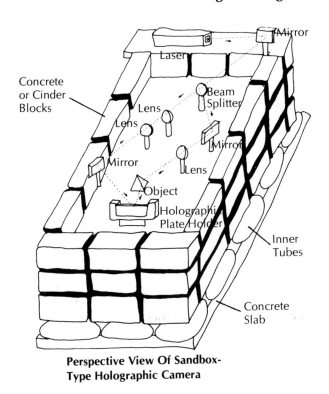

Perspective View Of Sandbox-Type Holographic Camera

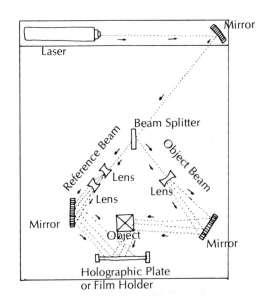

Top View Of Holographic Camera Set-Up

(Arrows indicate direction of laser light; dots indicate path of laser light.)

the holographic plate. This interference pattern is simply a wave pattern of light reflected from the subject. It contains *all* the information about how the subject is illuminated. Unlike photography in which a single image of the subject is recorded, a hologram contains a record of the many submicroscopic interference patterns that visually describe the illuminated subject completely. In this way, depth and parallax are preserved, and the hologram remains a record of the three-dimensionality of the subject.

Making a hologram consists of first dividing a single laser beam into two parts. One part of the laser beam, which illuminates the subject or object, is called the *object beam*. The other part of the laser beam, which is simply directed onto the holographic plate, is called the *reference beam*. Since the reference beam contains no information about the subject, its only function is to act as a standard of intensity to which the object beam can refer.

When a holographic plate or film is exposed to an object beam and a reference beam simultaneously, an interference pattern of the interaction is recorded. This re-

Recording A Hologram

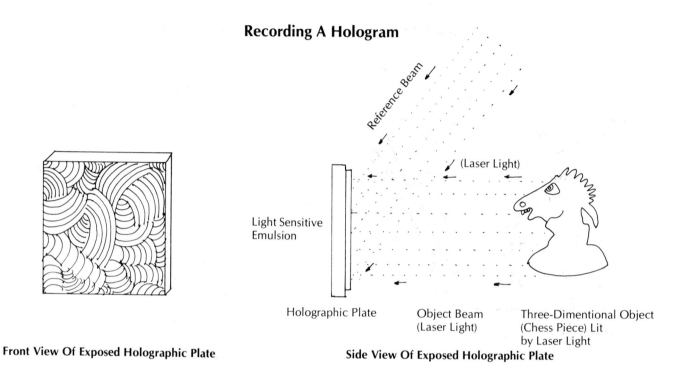

Light Sensitive Emulsion

Reference Beam

(Laser Light)

Holographic Plate

Object Beam (Laser Light)

Three-Dimentional Object (Chess Piece) Lit by Laser Light

Front View Of Exposed Holographic Plate

Side View Of Exposed Holographic Plate

Making a hologram is basically recording the interference pattern that occurs when a reference beam and an object beam interact with a light-sensitive emulsion (the holographic plate).

corded interference pattern is called a hologram. The entire setup, including laser, shutter, lenses, mirrors, film, and the isolation table that houses all the components, is collectively called a holographic camera.

The exposed holographic plate is developed the same way as in ordinary black-and-white photography. The fully processed hologram is then illuminated by laser light that approaches the plate from the same angle as the original reference beam. The hologram, which is the recorded interference pattern, then divides the incoming reference-beam light into an exact reproduction of the light from the object beam. The reproduction of the object-beam light becomes visible as an exact, three-dimensional replication of the subject of the hologram. In other words, the holographic image of the subject literally cannot be distinguished from the real and material three-dimensional object.

Viewing A Hologram

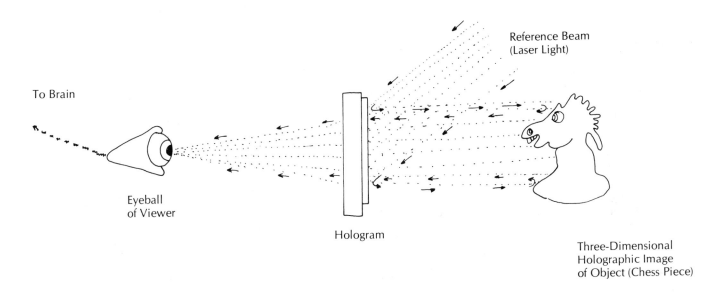

When a hologram that has been processed like an ordinary black-and-white photograph is re-illuminated by laser light, the hologram itself divides the reference beam into an exact reproduction of the object beam light. This light (the holographic image) then enters the eye of the viewer.

2-2 THE COHERENCE OF LIGHT THROUGH SPACE AND TIME

The light used in making holograms must be relatively coherent. In other words, this light must be *monochromatic:* that is, light of only one color, or a single wavelength. In addition, the light must be capable of interference. Most commonly, this light is laser light.

According to the wave description of light, *coherence* refers to light in which two or more light waves stick together or cohere to one another in a group in their passage through space and time. In these terms, coherent light consists of light waves of the same wavelength moving together and in the same direction. Noncoherent light is light of different wavelengths moving in different directions through space and time.

The Coherence Of Light

According To The Wave Description Of Light:

Same Wavelength of Light Moving through Space and Time Together in the Same Direction

Coherence

Different Wavelengths of Light Moving through Space and Time in Different Directions

Non-Coherence

According To The Photon Description Of Light:

Wave Packets (or Photons) with Equal Energy Values Moving through Space and Time Together

Coherence

Wave Packets (or Photons) with Different Energy Values Moving through Space and Time in Different Directions

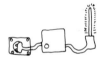

Non-Coherence

Sources Of Relatively Noncoherent Light

The Sun

Mercury Arc Lamp

Tungsten-Filament Light Bulb

Neon Tubes

According to the quantum description of light, coherent light consists of photons of a single frequency that cohere to one another as they oscillate. Since photons are actually packets of energy traveling in wave form, they are also called *wave packets* in this case. The most coherent state of light refers to these wave packets of equal energy traveling together in the same direction. Noncoherence refers to wave packets of different frequencies and different energy values moving through space and time in different directions.

Ordinary sources of light, such as incandescent and fluorescent lamps, are noncoherent, or less accurately called, "incoherent." Noncoherent sources of light include the sun, fire, light bulbs, electric arc lamps, and neon tubes. Unlike laser light, the light from these sources has neither temporal nor spatial coherence.

Temporal coherence is the coherence of light with respect to time. When the frequency and wavelength of light remain constant in relation to the amplitude, temporal coherence exists. In other words, as time passes, both the wavelength and the intensity of the light do not change. This type of light is called monochromatic and is characterized by a more or less single frequency of unvarying intensity.

Ordinary sources do not produce monochromatic, or single-frequency, light. Instead, they emit light that includes many different frequencies, all vibrating out of phase with one another. The sun, for example, emits electromagnetic energy in all visible frequencies as well as in all the invisible ones.

In order to simulate monochromatic light with an ordinary light source, a mercury arc lamp with a color filter can be used. The filter allows only the desired wavelength of light to pass through, producing what can be considered monochromatic light for most purposes. Laser light, however, is really the ideal light source for holography since it emits a single frequency, or very nearly so, at an unchanging intensity.

In terms of the wave description of light, temporal coherence refers to two or more light waves of the same wavelength whose phase difference remains constant. When two light waves are in phase, a constant phase difference means no difference in phase, which is the same as a constant phase difference of zero. Graphically, two light waves that are in phase can be represented by two sine curves whose maximum and minimum points correspond exactly to one another.

When two light waves of the same wavelength are out of phase, there can also be a constant phase difference. Since the maxima and minima of the two waves are a constant distance apart, these two out-of-phase light waves are also temporally coherent. Graphically, this situation can be represented by two sine curves with equal amplitudes whose maxima and minima don't correspond exactly to one another.

In holography, temporal coherence is usually measured in terms of the coherence length of the laser. Coherence length is simply the actual length of the laser tube. This distance represents the length of a continuous and uninterrupted series of coherent light waves that emit from the laser. Because of this relation, coherence length of a laser actually sets a limit on the size of the subject. If the depth of the subject of the hologram is longer than the coherence length, the light that illuminates the subject will not be temporally coherent, so no image will be recorded.

Temporal coherence depends on a constant phase difference between any set of light waves, whether in phase or out of phase.

Spatial coherence refers to light waves that are in phase along the direction of movement and along any direction perpendicular to the direction of movement. Spatially coherent light is light that remains in phase in its movement through space.

Spatial coherence is the coherence of light with respect to space. When the amplitude of light varies directly as a sine wave, some degree of spatial coherence exists. In other words, light that is spatially coherent constantly and continuously oscillates through space and time in the form of a sine wave.

In a more rigorous definition, spatial coherence specifically refers to light waves that are in phase relative to the direction of movement and also relative to any direction perpendicular to the direction of their movement. Graphically, this can be described by two sine curves representing two light waves in phase so that maxima, zero points, and minima of the two waves coincide exactly. This situation satisfies the condition for the temporal coherence of light when there is no phase difference between light waves.

In addition, the two light waves must be in phase so that the distance between the two waves remains constant. This lateral, or sideways, equal distance between two light waves fulfills the condition for lateral coherence. When two light waves that are in phase exhibit both temporal coherence and lateral coherence, the light is then spatially coherent.

Sources of spatially coherent light are commonly called *point sources* of light. Visually, a point source appears to emit a point or actually a small spot of light. Or-dinary sources of light such as light bulbs are not considered point sources since the light they emit is not spatially coherent. Instead, light from a light bulb includes many frequencies at different intensities.

In order to simulate a point source with an ordinary source of light, a light bulb with a pinhole can be used. When placed next to a light bulb that emits light, a pinhole spatially filters the otherwise broad band of light into a point of light. When light passes through a pinhole, it becomes spatially coherent to some degree. At its simplest, a pinhole consists of a piece of cardboard with a literal pinhole or a small circular hole.

The laser, however, is a source of light that produces monochromatic light which is also spatially coherent without the use of a pinhole. The laser, in effect, is a point source of light.

The motion of the light emitted by a point source can be visualized by imagining an ever-expanding cone of light that spreads out in all directions in its movement through space and time. This expanding cone of light in three dimensions represents the history of the motion of light as well as its expansion.

At any instant of time, the motion of light emitted from a point source can be described three-dimensionally by a series of expanding concentric spheres. The surface of each sphere represents the position

The Coherence Of Light In Time
(Temporal Coherence)

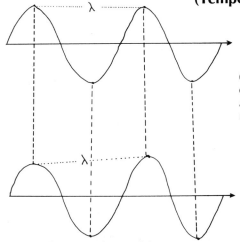

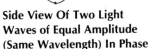

(Arrow designates the direction of light movement as well as the light movement through time.)

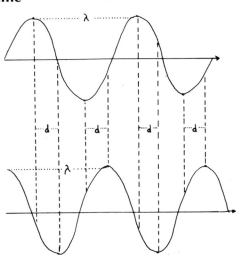

Side View Of Two Light Waves of Equal Amplitude (Same Wavelength) In Phase

(There is no phase difference; that is, there is a constant phase difference of zero.)

Side View Of Two Light Waves Of Equal Amplitude (Same Wavelength) Out Of Phase

There is a constant phase difference since maxima and minima are a constant distance [d] apart.

The Coherence Of Light In Space
(Spatial Coherence)

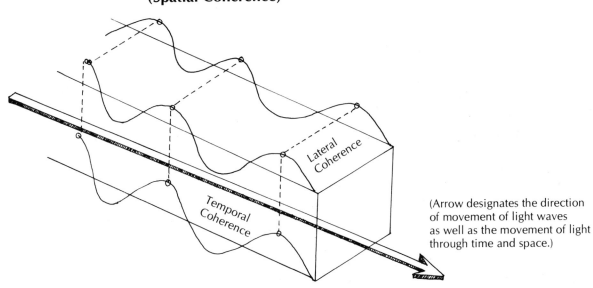

(Arrow designates the direction of movement of light waves as well as the movement of light through time and space.)

Temporal Coherence And Spatial Coherence In Graphic Form

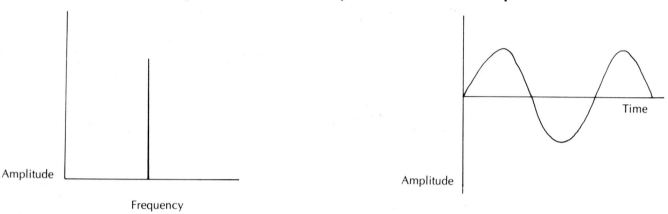

Amplitude

Frequency

Temporal Coherence Graph

Amplitude

Time

Spatial Coherence Graph

Temporal coherence refers to light, the frequency (and wavelength) of which remains constant in relation to its amplitude; in other words, monochromatic light of an unvarying intensity. Spatial coherence refers to light, the amplitude of which varies directly as a sine wave in relation to time; in other words, light that continuously oscillates, or vibrates, in the form of a sine wave (above).

Sources Of Spatially Coherent Light
(Point Sources Of Light)

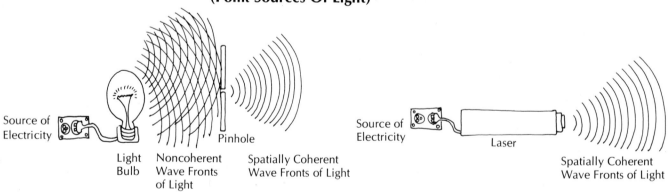

Source of
Electricity

Light
Bulb

Noncoherent
Wave Fronts
of Light

Pinhole

Spatially Coherent
Wave Fronts of Light

Source of
Electricity

Laser

Spatially Coherent
Wave Fronts of Light

Side View Of Ordinary Light Source

Side View Of Laser

In diagram below, an ordinary source of light emits light of many frequencies at different intensities. A pinhole alters this light, producing more spatially coherent wave fronts of light (left). A laser emits light of a single frequency (monochromatic light) which is spatially coherent without the use of a pinhole (right). In effect, a laser is a point source of light.

The Motion Of Light Through Space-Time
(Movement Of Light In Four Dimensions)

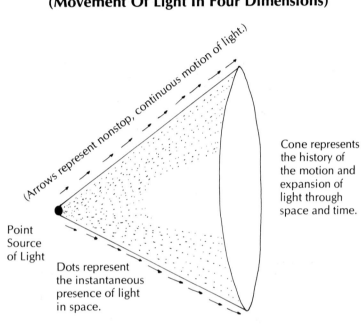

The motion of light through space and time can be simply represented by an ever-expanding cone of light. Due to the limitations of a two-dimensional drawing, one of the spatial dimensions has been excluded. Also, it should be noted that this is a static pictorial description of energy in motion.

of light energy at any given time and also constitutes what's called a *spherical wave front* of light. The point source of light is in the exact center of all the expanding concentric spheres. A physical model of the motion of light consists of a balloon being blown up so that the surface area is constantly expanding. The surface area of such a balloon represents the constantly expanding spherical wave fronts of light.

Two-dimensionally, spherical wave fronts can be described by bisecting, or cutting in half, the series of concentric spheres. In this model, the resulting over-

head view of the wave fronts consists of a series of concentric circles with the point source in the center. Each circle, spaced exactly one wavelength apart, represents a spherical wave front emitting from a point source.

When a sample is taken from a series of spherical wave fronts that are far removed from the source of light, the spherical wave fronts tend to flatten out. Because of this flattening-out effect, the spherical wave fronts can be considered to be *plane wave fronts*. These plane wave fronts are represented by parallel lines spaced one

The Motion Of Light Through Space
(Movement Of Light In Three Dimensions)

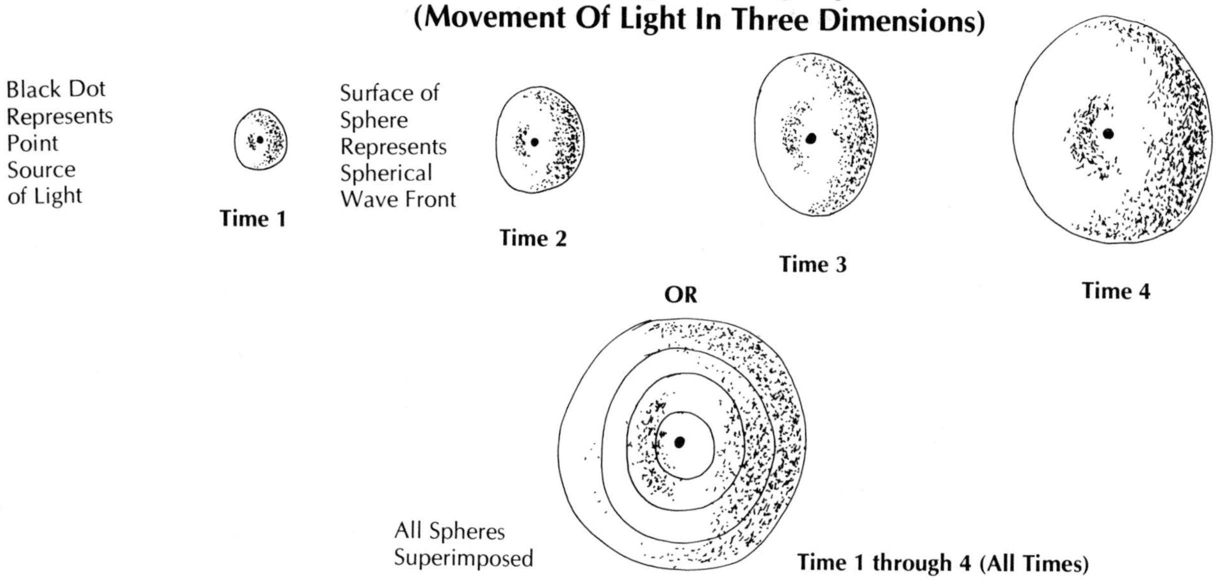

Black Dot
Represents
Point
Source
of Light

Time 1

Surface of
Sphere
Represents
Spherical
Wave Front

Time 2

Time 3

Time 4

OR

All Spheres
Superimposed

Time 1 through 4 (All Times)

The motion of light through space at any instant in time can be represented by a sequence of spheres, each sphere larger than the preceding one. The surface of each sphere represents a spherical wave front of light at any given time. The point source of light is in the center of each sphere. When the instants of Time 1 through Time 4 are pictured together, a series of ever-expanding concentric spheres simply illustrates the motion of light during all times.

When a series of concentric spheres (each sphere surface representing a spherical wave front) is bisected or cut in half by a plane (a two-dimensional surface), the resulting overhead view of wave fronts consists of a series of concentric circles. The point source of light is in the center and each circle is spaced one wavelength of light apart (above).

When a sample is taken from a series of spherical wave fronts (represented by concentric circles) far removed from the source of light, a series of plane wave fronts is formed. These plane wave fronts can be represented by parallel lines spaced a wavelength apart (below).

wavelength apart. Each parallel line is a plane wave front of what is called a *plane electromagnetic wave.*

Two-dimensionally, a plane electromagnetic wave, or simply, a plane electric wave, can be described in terms of an overhead view and a side view. An overhead view of plane wave fronts of light can be represented by parallel lines that are

spaced a wavelength apart. Each parallel line, in this case, represents a plane wave front of light.

The side view of a plane electric wave consists of a sine wave. Corresponding to the overhead view of the plane wave fronts of light, each wave front is formed by a plane that passes through each maximum and each minimum of the sine curve. The

The Motion Of Light Through Space
(Two-Dimensional Representation Of The Movement Of Light)

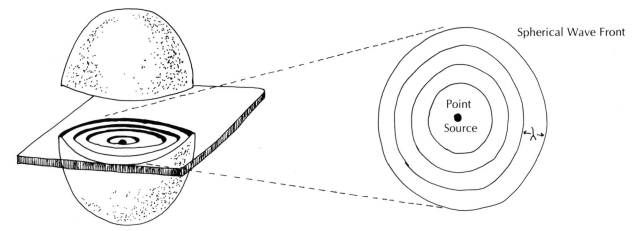

Spherical Wave Front

Side View Of A Three-Dimensional Representation Of Light Motion (A Series Of Concentric Spheres) Divided Into Two Hemispheres

Top View Of A Two-Dimensional Representation Of Light Motion (A Series Of Concentric Circles That Represents Spherical Fronts of Light)

Two-Dimensional Representation Of Spherical Wave Fronts And Plane Wave Fronts

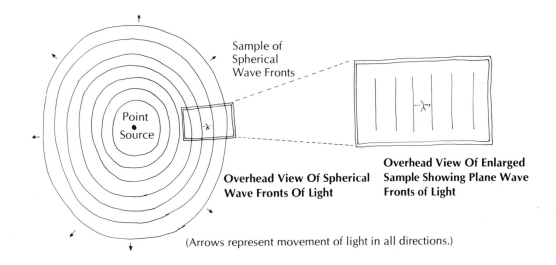

Sample of Spherical Wave Fronts

Overhead View Of Spherical Wave Fronts Of Light

Overhead View Of Enlarged Sample Showing Plane Wave Fronts of Light

(Arrows represent movement of light in all directions.)

Two-Dimensional Representation of Plane Wave Fronts Of Light And The Light Wave

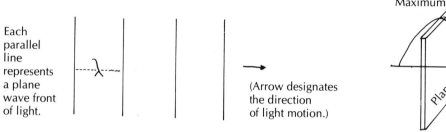

Each parallel line represents a plane wave front of light.

(Arrow designates the direction of light motion.)

Overhead View Of Plane Wave Fronts Of Light

Side View Of Plane Electric Wave

Each plane wave front of light corresponds to each plane that passes through each maximum and minimum of a sine wave. Each parallel line represents one wave front of light. The sine wave represents a simple light wave or more specifically, a plane electric wave.

side view and overhead view combined constitute a static, two-dimensional map of light in motion.

Because of its spatial and temporal coherence, laser light is ideally suited as the source of illumination in making holograms. However, the use of a laser in viewing holograms is completely impractical because of the expense of purchasing and maintaining it. For this reason, light bulbs are much more appropriate.

2-3 INTERFERENCE

At its simplest, interference refers to the interaction or the actual contact of two or more coherent waves of light. The coherent light, in this case, is composed of monochromatic light waves and is produced by a single continuously emitting source. The source of light is, of course, the laser.

Wherever two or more light waves from the same laser interfere with one another, an interference pattern is formed in space. When this space of interference is intersected by a two-dimensional plane, such as a card, for example, the interference pattern becomes visible. The interference pattern consists either of concentric alternately light and dark circles or alternately light and dark bands called *interference fringes*.

Since the interference fringes appear to be stationary, or standing in place, the interference pattern is also called a *standing wave pattern*. The interference of light waves appears as a stationary pattern despite the unimaginably high frequency of the vibration of light several trillion cycles per second. The interference pattern appears not to be moving because the human eye is not sensitive enough to register the individual vibrations of light in motion. Instead, these extremely small but continuous oscillations of light are blurred so that the human eye perceives an apparently stationary interference pattern, or standing wave pattern.

An interference pattern is visible not only when coherent light waves interact, but also when noncoherent light such as sunlight interferes with itself. The interference of coherent light, however, is much simpler to describe since light of only a single wavelength is involved. When light waves of different wavelengths, different

So-Called Constructive Interference Of Light
(In Terms Of Amplitude Addition Of Two Waves)

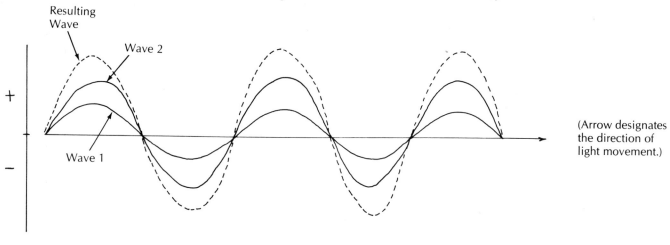

When two light waves of the same wavelength in phase with each other interfere, so-called constructive interference occurs. This can be represented graphically by the superimposition of two sine curves of different amplitudes, so that amplitude addition of the two waves of light takes place. The interference of two light waves results in a single light wave, the amplitude of which is the sum of the two amplitudes.

amplitudes, and different phase relations interfere with each other, a much more complex set of interference fringes is produced. For this reason, the least complicated example of light-wave interference involves two light waves of the same wavelength with equal amplitudes.

An interference pattern is a relatively gross visible manifestation of a sub-microscopic physical event—the interference of coherent light waves. According to the wave description of light, interference refers to the change in the overall intensity of light that takes place when at least two light waves are superimposed on each other. In the simplest instance, two waves of light in phase with each other meet in space, thereby producing another light wave. The resulting light wave is created by the addition of the amplitude values of the original two waves.

When the two amplitudes are added, maximum to maximum and minimum to minimum, light of a greater amplitude is produced. Greater amplitude means an intensity of light greater than that produced

So-Called Destructive Interference Of Light
(In Terms Of Amplitude Addition Of Two Waves)

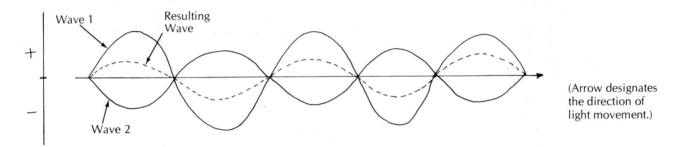

So-called destructive interference takes place when two light waves of the same wavelength, but out of phase, (above, 180° out of phase) interfere with each other. The resulting amplitude addition produces a wave with a decreased total amplitude.

by a single wave alone. This superimposition of light waves in the form of amplitude addition is commonly referred to as *constructive interference.*

Constructive interference accounts for the light fringes in an interference pattern. Addition of the amplitudes of two light waves that are in phase represents a greater total intensity of light. A greater overall light intensity produces the light fringes of an interference pattern.

When two light waves that are 180° out of phase interfere with each other, a light wave with a decreased amplitude is produced. The decreased amplitude results from the addition of amplitude values of opposite polarity. In other words, the addition of positive (+) to negative (−) amplitude values results in a decreased total amplitude corresponding to a lower intensity of light. When the resulting intensity of light is lower than either intensity alone, the event is called *destructive interference.*

Destructive interference can also occur between two light waves with equal

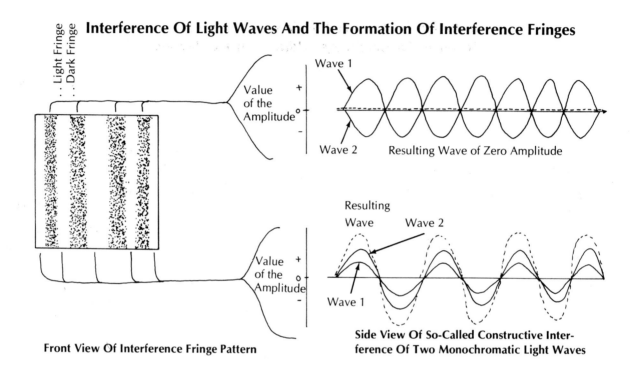

Front View Of Interference Fringe Pattern

Side View Of So-Called Constructive Interference Of Two Monochromatic Light Waves

When two monochromatic light waves that are 180° out of phase interfere with each other, so-called destructive interference takes place. The dark fringes signify absence of light energy. Likewise, so-called constructive interference is responsible for the light fringes.

amplitudes and 180° out of phase with each other. When the two equal amplitudes are added, maximum to minimum, and so on, the result is a zero intensity of light, which, in a word, is darkness. In this case, destructive interference accounts for the absence of light—the dark fringes in an interference pattern.

These expressions, "constructive interference" and "destructive interference," are at least misleading if not outright confusing. The interference of light has absolutely nothing to do with either construction or destruction, particularly of

light itself. Even in the interference of two light waves with equal amplitudes and 180° out of phase with each other, light is not destroyed but is simply displaced to another location. When such a displacement of light energy occurs, the result is darkness.

The displacement of light is also commonly (and incorrectly) called *cancellation*. The term cancellation is supposed to imply the balancing out of amplitude values of opposite polarities. However, because of its use in post and box offices, the word cancellation should be avoided in

Making A Hologram By Amplitude Division

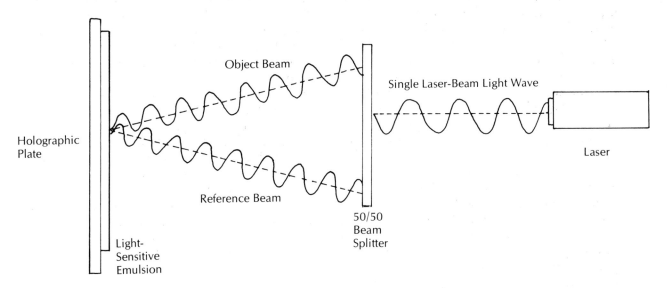

Side View Of Amplitude Division And Recombination

When a single laser beam passes through a beam splitter, two beams of light are produced. Each of these two laser beams consists of light waves whose amplitude is half the amplitude of the single laser-beam light wave. When two beams of light (the object beam and the reference beam) interfere, their interference pattern can be recorded on a holographic plate. The recorded interference pattern is the hologram itself.

describing interference. The absence of energy, which is darkness, is not a cancellation, nor is it an extended run, for that matter.

Making a hologram means recording the intensity of the interference of two coherent beams of light. By using a beam splitter, the light of a single laser beam is divided into two beams. The division of light in effect divides the amplitude of the single laser beam into two parts. Each of the resulting two beams of light, then, has half the intensity of the original single light beam. After the two laser beams have both traveled the same distance, their interference can be recorded on a photosensitive material—the holographic plate or film. The recorded pattern of interference between the two beams of laser light is the hologram itself. The two beams of laser light are called the *object beam* and the *reference beam*. The division of this light and its reunion on the holographic film basically describes the process of holography.

Another example of the interference of coherent light waves is the so-called *speckle pattern* that appears at the operation of any laser. Speckle pattern refers to the scintillating or sparkling pattern of speckles, which are very small spots of light and darkness. This stationary pattern of speckles is caused by the highly coherent laser light which interferes with itself on the surface of the illuminated subject. The graininess of the pattern is due to the interference between light waves that have been reflected and diffused by the subject. Also, for this reason, holograms illuminated by laser light have a grainy appearance.

Since laser light is highly coherent or constant in time, the speckle pattern appears to be stationary. This appearance, however, can be averaged out by changing the position of the viewer's eye in relation to the speckles. Averaging out the speckle pattern means scanning the holographic image with the eye to counteract the grainy appearance.

Making a hologram consists of recording the interference pattern produced by the interference of two beams of coherent light, the object beam and the reference beam. The recorded interference pattern, the hologram, contains information about the intensity of both light beams as well as the phase relation between the object beam and the reference beam. In other words, because a hologram is recorded using relatively coherent light, the relationship of phase is preserved as part of the interference pattern.

In photography, however, only the intensity of light is recorded. Since noncoherent light such as sunlight is used, the phase relation between different light waves is lost. Noncoherent light contains not only all wavelengths of light, but also all possible phase relations between light waves. Because the phase relation is constantly changing, no information about the phase can be recorded in photography.

Since the recorded interference pattern, the hologram, is capable of retaining information about the phase as well as the intensity of light, the holographic image likewise retains the "real" and three-dimensional appearance of the subject.

2-4 DIFFRACTION

In general, diffraction refers to the observable effects that appear when light or any other electromagnetic radiation encounters an obstacle in its path. When this occurs, light seems to bend around the edge of the obstacle instead of continuing to travel along an undeviated course.

When light is broken up, or diffracted, by some obstacle in its path, the visible result is commonly called a *diffraction pattern*. A diffraction pattern appears when light passes near the edge of an opaque object—a straight edge such as a razor blade, for instance. A diffraction pattern is also formed when light passes through an aperture, an opening like a narrow rectangular slit, or a circular aperture like a pinhole.

A rectangular aperture, which is called a slit, produces a diffraction pattern of alternately light and dark bands. A circular aperture or pinhole produces a set of alternately light and dark concentric circular rings. The interaction of light from several diffraction patterns is called an interference pattern. An interference pattern is formed by multiple diffraction from more than one slit, aperture, or source of light.

The diffraction pattern produced by light passing through a single rectangular aperture is called a single-slit diffraction pattern. The appearance of this pattern can be explained by considering the diffracted light that emerges from the aperture to be composed of secondary wave fronts of light. The so-called secondary wave fronts, which radiate in all directions, are emitted by secondary point sources. These secondary sources of light are located at each and every point on the original diffracted wave front. Since constructive and destructive interference occur naturally between all secondary wave fronts, a simple diffraction pattern of alternately light and dark fringes is produced.

A single-slit diffraction pattern consists of a centrally located, relatively broad and bright fringe that corresponds to light

The Visual Effects Of Diffraction
(Diffraction Patterns)

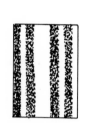

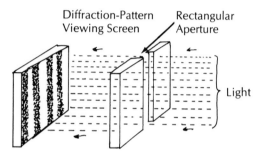

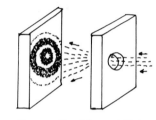

Diffraction-Pattern Viewing Screen   Rectangular Aperture

Light

Diffraction-Pattern Viewing Screen

Front View of Diffraction Pattern

Perspective View Of Light Passing Through A Rectangular Aperture (A Slit)

Front View of Diffraction Pattern

Perspective View of Light Passing through a Circular Aperture (a pinhole)

(Arrows signify the direction of light movement.)

which simply passes through the aperture without diffraction. Also, this pattern consists of relatively narrower light and dark fringes on either side of the central light fringe. The narrower fringes correspond to the mutual interference of all the secondary wave fronts.

When light passes through two narrow rectangular apertures whose widths each approximate a wavelength of light, a double-slit diffraction pattern is produced. This pattern is similar to the single-slit diffraction pattern except that the centrally

located light fringe is broken up into several narrower light and dark fringes. Each light fringe is also called a maximum because it corresponds to a maximum intensity of light. Likewise, each dark fringe is called a minimum.

A double-slit diffraction pattern is simply the result of two kinds of interference. First, interference takes place between secondary wave fronts that arise at each individual slit. This interference pattern is simultaneously superimposed on the pattern which is caused by interference

The Formation Of A Single-Slit Diffraction Pattern

Light
Fringe

Dark
Fringe

Viewing Screen

Graph of Relative
Light Intensity
at Viewing Screen

In relation to the intensity of light:
maxima correspond to light fringes;
minima correspond to dark fringes.

(Arrows indicate
the direction of
light movement.)

Diffracted Spherical
Wave Fronts of
Light (So-Called
Secondary Wave Fronts)

Plane Wave Fronts of
Monochromatic (Single-
Wavelength) Light

Rectangular Aperture
(a Single Slit)

**Front View Of Viewing
Screen (Single-Slit
Diffraction Pattern**

**Side View Of Light Passing
Through A Rectangular
Aperture (A Single Slit)
Onto A Viewing Screen**

between wave fronts from both slits to-gether. Because terminology in optics is used so loosely and even more randomly devised, a double-slit diffraction pattern is also commonly called an interference pattern.

When the aperture through which light passes consists of many or multiple parallel slits of identical width and separation, the aperture is called a *diffraction grating*. The diffraction pattern from such a grating consists of a large number of much narrower interference fringes. As the

number of slits is increased, the interference fringes become narrower and sharper until they appear not as broad fringes but as narrow microscopic lines.

Ordinarily, diffraction gratings are made by so-called ruling machines or ruling engines. With a diamond point, the ruling machines make extremely fine grooves on either glass plate, speculum metal (which is a hard, mirrorlike alloy of copper and tin) or on evaporated layers of aluminum. The gratings made by such machines yield tens of thousands of

The Formation Of A Double-Slit Diffraction Pattern

Dark Fringe

Light Fringe

Viewing Screen

Graph of Relative Light Intensity at Viewing Screen

(Arrows indicate the direction of light movement.)

Plane Wave Fronts of Monochromatic Light

Diffracted Spherical Wave Fronts of Light

Two Rectangular Apertures (a Double Slit)

**Front View Of Viewing Screen
(Double-Slit Diffraction Pattern)**

**Side View Of Light Passing Through
Two Rectangular Apertures
(A Double Slit) Onto A Viewing Screen**

grooves per inch.

When grooves are etched in glass, the incoming light is scattered in the same way as it is by the opaque areas between the slits. The unruled glass allows light to pass through as it does through the slit itself. Since light is transmitted through the unruled glass, this type of grating is called a transmission grating.

When the grooves are etched in aluminum, the light that strikes the unruled aluminum surface is reflected back into the direction of the incoming light. This type of grating is called a reflection grating.

When the mirrored aluminum surface is concave and spherical, the grating is called a concave grating. The concave grating not only diffracts or reflects incoming light, but focuses it as well. Because of this focusing ability, the concave grating can be used to diffract ultraviolet frequencies of light that are not transmitted by glass as well as to compare and measure different wavelengths of light.

Holograms can be considered to be diffraction gratings since they also diffract light. Diffraction in a hologram occurs throughout the depth of the emulsion. Inside the emulsion, the internal boundaries

The Formation Of A Multiple-Slit Diffraction Pattern
(The Diffraction Grating)

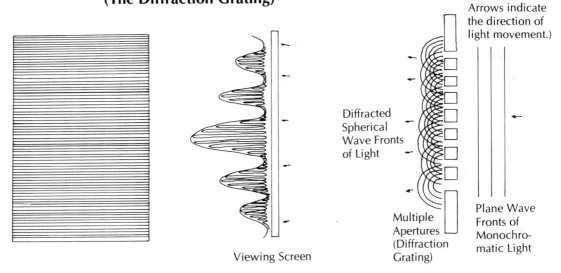

Front View Of Viewing Screen
Dark and light interference fringes are
sharp and narrow lines.

Viewing Screen

Arrows indicate the direction of light movement.)

Diffracted Spherical Wave Fronts of Light

Multiple Apertures (Diffraction Grating)

Plane Wave Fronts of Monochromatic Light

Side View Of Light Passing
Through Multiple Rectangular Apertures
(A Diffraction Grating) Onto A Viewing Screen

that guide incoming light through the hologram are the planes of silver deposited at the time the hologram is developed. These planes of deposited silver act like perfect mirrors in reflecting or diffracting the light.

The planes of deposited silver atoms correspond exactly to the planes of recorded light interference fringes. When the planes of silver are stacked more or less perpendicular to the length of the holographic plate, the incoming light is transmitted through the hologram. A hologram that diffracts light by transmission is called a *transmission hologram*.

When the planes of deposited silver grains throughout the emulsion lie parallel to the plane of the holographic plate, incoming light is reflected back outward. In other words, the holographic image is seen by reflection. A hologram that diffracts light by reflection is called a *reflection hologram*.

The simplest holographic grating can be made by recording the interference of two beams of relatively coherent light—a reference beam and an object beam. The reference beam can be considered to be a series of plane and parallel wave fronts of light. The subject of the ob-

Types Of Diffraction Gratings

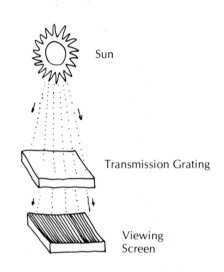

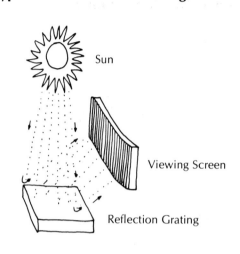

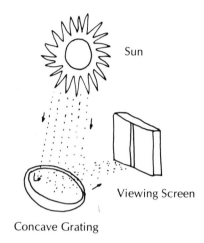

Side View Of Sunlight Passing Through A Transmission Grating

Side View Of Sunlight Diffracted By A Reflection Grating

Side View Of Sunlight Diffracted And Focused By A Concave Grating

ject beam is a single point of light. Since the point of light, which is the object, is at a great enough distance from the holographic plate, it can also be considered a source of a series of plane and parallel wave fronts of light.

At the plane of the holographic plate, each object wave front interferes with each reference wave front of light. Since constructive and destructive interference occur, respectively light and dark interference fringes are recorded throughout the emulsion of the holographic plate.

The hologram itself consists of many superimposed diffraction gratings. One grating is formed for each plane of interference, that is, each interference fringe corresponds to one diffraction grating.

After photographic processing, the hologram is re-illuminated by a reference beam of plane wave fronts of coherent light. The diffraction grating, which is the hologram, diffracts part of the incoming light into the direction of the object beam. This diffracted light consists of plane and parallel wave fronts of light which are exactly identical to the wave fronts of the original object beam. The single point of light that is the object beam is reproduced exactly.

Holograms As Diffraction Gratings

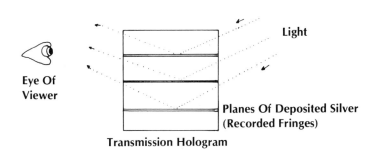

Transmission Hologram

Side View Of A Sample Of
A Transmission Hologram

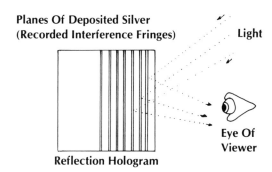

Reflection Hologram

Side View Of A Sample Of
A Reflection Hologram

Light that enters a hologram or a diffraction grating is diffracted according to the so-called *grating equation*. The grating equation, also called Bragg's law, is as follows:

$$2\theta \approx \frac{\lambda}{d}$$

In this relation, 2θ is the angle between incident or incoming light and diffracted or exiting light, λ is the wavelength of incident light, and d is the width of each interference fringe. The grating equation is read: Two times the angle theta is approximately equal to the wavelength of light divided by the width between interference fringes.

The usefulness of the grating equation lies in the fact that if the wavelength of light and the angle between reference and object beams are known, the spacing, or width, between interference fringes can be calculated. Also, using this relation, the angle at which the hologram diffracts light most efficiently can be determined. The grating equation is equally applicable to all forms of electromagnetic radiation, though light and X-rays are most commonly used in studies of diffraction.

The relative scale of these submicroscopic physical events should be kept in mind. The interference fringes are re-

The Grating Equation (Bragg's Law—$2\theta \approx \frac{\lambda}{d}$)

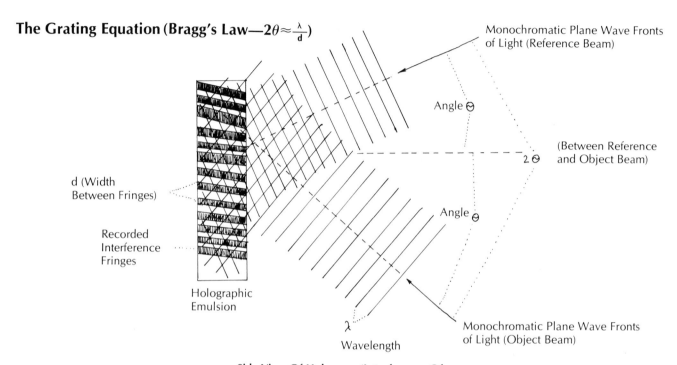

**Side View Of Hologram (Interference Of
Two Monochromatic Sets Of Wave Fronts)**

The grating equation is read: two times the angle theta (2θ) is approximately equal to the
wavelength of light divided by the width between interference fringes (λ).

corded throughout an emulsion that is typically 6 to 15 nanometers wide. A nanometer is one millionth of a millimeter. The silver-halide grains found in the high-resolution, fine-grain emulsions used in holography are usually less than one tenth of a nanometer wide.

The diffraction pattern that appears when light passes through a single circular aperture consists of concentric circular alternately light and dark rings or circular interference fringes. These circular fringes are also called Fresnel zones (the "s" in Fresnel is silent).

When a helium-neon laser, which emits red light, is used, the interference fringes appear as alternately red and black concentric rings. If the Fresnel zones are recorded by photographing the diffraction pattern, the photographic transparency is called a Fresnel zone plate, or simply, a zone plate. A zone plate is a diffraction grating that acts as both a positive lens as well as a negative lens. In other words, having two focal points, a zone plate has the ability to focus light in front and in back of itself.

A hologram can be made of the Fresnel zones of a single point of light that serves as the object. This is done by divid-

The Formation Of A Fresnel-Zone Diffraction Pattern

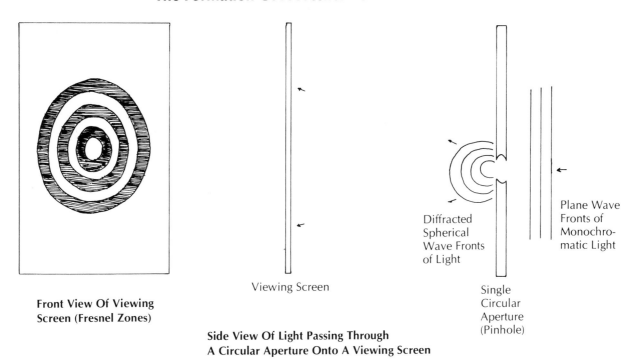

**Front View Of Viewing
Screen (Fresnel Zones)**

Viewing Screen

**Side View Of Light Passing Through
A Circular Aperture Onto A Viewing Screen**

Diffracted
Spherical
Wave Fronts
of Light

Single
Circular
Aperture
(Pinhole)

Plane Wave
Fronts of
Monochro-
matic Light

ing a laser beam into a reference beam, which illuminates the entire holographic plate, and an object beam consisting of light that passes through a pinhole. The resulting hologram is also called a zone plate and it also behaves like a diffraction grating capable of focusing light. In addition, if a holographic zone plate is illuminated by a reference beam, a replication of the point source, the object beam, will be produced.

Just as a zone-plate hologram is capable of reproducing the wave fronts of light from a single point source, so a hologram of an object is capable of reproducing the wave fronts of light from an entire three-dimensional object. Since the light reflected from a three-dimensional object is simply a spatial grouping of many point sources of light, a hologram of a three-dimensional object is simply a set of as many superimposed zone plates. When the processed hologram is re-illuminated by the reference beam, each individual zone-plate hologram reproduces its own particular set of wave fronts. When all the sets of wave fronts are seen together at once, a holographic image, or a three-dimensional replication of all the object light, becomes apparent.

2-5 WAVE-FRONT RECONSTRUCTION

The method of holography is basically a two-step process. The first step consists of recording the coherent wave fronts of light reflected by an illuminated object. In order to record these wave fronts, holographic film is exposed to the interference pattern produced by an object beam and a reference beam. The record of the interference between the two sets of wave fronts is the hologram itself.

The second step of the process consists of reproducing the object-beam wave fronts from the hologram. Coherent light, which acts as the reference beam, re-illuminates the hologram from the same angle as it did in the recording. The re-illuminated hologram in turn diffracts light into a set of wave fronts identical with the original object beam. These object-beam wave fronts collectively appear as the holographic image. Generally, this is known as *wave-front reconstruction*. The choice of the term "wave-front reconstruction" is literally misleading, since no construction or building of wave fronts actually takes place. What happens, however, is that the hologram itself diffracts wave fronts of light that are exactly the same as the wave fronts of object-beam light during the recording.

In order to make a hologram, the light from a single laser beam is divided by a beam splitter into two beams of light, the object beam and the reference beam. A beam splitter is simply a piece of ordinary glass that allows about 90 per cent of the light through and deflects about 10 per cent of the light to the side. The object beam is the laser light that illuminates the object and then is reflected from the object onto a holographic plate. The reference beam is unmodified laser light directed at the holographic plate.

An interference pattern is formed in the space where the two beams of laser light overlap. When a light-sensitive material such as a holographic plate is placed in the space where interference occurs, the exposure to the energy of light will form a preliminary record of an interference pattern. After the holographic plate is developed, stopped, and fixed like an ordinary black-and-white photograph, the recorded interference pattern becomes a hologram. The hologram then becomes capable of wave-front reconstruction and is able to produce a three-dimensional holographic image.

The word "hologram" is derived from the Greek words that mean a whole or complete description. A hologram is so named because it actually consists of a record of all the phase and amplitude information from the object. The completeness of the information is most striking because a hologram produces an image of light, a picture that is an actual three-dimensional copy of a real object. (The word "hologram," by the way, should not be confused with the word "holograph," which is a document written entirely in a person's own handwriting.)

Wave-front reconstruction occurs in the second step of the process when the hologram is re-illuminated by the original reference beam. The reference beam, which is just plain laser light, illuminates the

Types Of Holographic Images Produced By Wave-Front Reconstruction

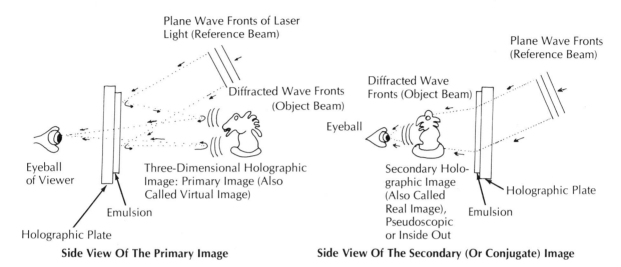

Side View Of The Primary Image

Side View Of The Secondary (Or Conjugate) Image

Plane wave fronts of light illuminate the hologram. The hologram diffracts object wave fronts that form the holographic image. The light from the holographic image enters the eye (left). Plane wave fronts of light illuminate the hologram. The hologram diffracts light into the secondary image. These diffracted wave fronts enter the eye of the viewer. Note the change of emulsion side of the holographic plate (right).

holographic plate from the same angle as when the hologram was made. The hologram itself, then, diffracts or scatters part of the reference beam into a reconstruction of the wave fronts of light as they were reflected by the object. These so-called reconstructed wave fronts yield a three-dimensional image that simply cannot be told apart from the real and material three-dimensional object. The depth as well as the parallax in the holographic image is exactly the same as in the object. Parallax, in this sense, allows the observer to look around the object and to see it from different angles.

When the observer looks through the holographic plate as if through a window, the holographic image is found precisely in the same volume of space as was the real object. This holographic image is usually called the *primary image,* or in the old terminology of optics, the *virtual image.* "Virtual," in this sense, means that there appears to be a source of light at the location of the image, but that this light is only due to laser light which is focused by the hologram into this exact space. The term "primary image" is preferred since none of the vagueness of the word "virtual" is attached to it.

The primary image is also called an *orthoscopic image.* Orthoscopic means that the image is relatively right side up and that its depth is correct or normal in relation to its length and width. In other words, an orthoscopic image is an image that is free of distortion.

When the opposite, or back, surface of the holographic plate is illuminated, another holographic image, the *secondary image*, is generated. This image, located in the space between the holographic plate and the observer's eye, is also called the *conjugate image.*

The secondary, or conjugate, image in this case is *pseudoscopic*. Pseudoscopic means that the image is correct in the two dimensions of length and width while the depth of the image is inverted, or turned inside out. The unusual effect of a pseudoscopic image can also be described as reversed parallax. The change in perspective as the observer moves from left to right in viewing the pseudoscopic image is the same as moving right to left in viewing the orthoscopic image. In addition, there is a reversed focus of the image itself. The reversed focus can be visualized by imagining a three-dimensional image of an object or a scene in which the background occupies the space of the foreground and the foreground appears where the background should be. This is difficult to imagine, but the background unexpectedly obscures, or covers, the foreground so that the mind responds with the impression that the image is inside out.

Interference and the subsequent reconstruction of wave fronts can be graphically described by the simple interference of two series of plane wave fronts. One series of plane wave fronts represents the reference beam. The reference beam is considered a series of plane electric waves because the laser light is collimated. In other words, when laser light passes through a collimating lens, the wave fronts of light become parallel to one another.

The object beam consists of light from an illuminated object, light that is scattered to every point on the holographic plate. Since the illuminated object is actually a grouping of points of light, the object beam is more accurately a set of spherical wave fronts. For the purpose of simplicity and demonstration, however, the object beam is considered to be composed of a single object point that radiates a series of plane wave fronts. Interference occurs between two sets of plane wave fronts in the simplest instance.

The interference of two plane waves or two sets of plane wave fronts can be represented by two sets of intersecting parallel lines. Each parallel line represents a single plane wave front. Also, each parallel line corresponds either to a maximum or a minimum value of an AC sine wave. In other words, each parallel line represents a single amplitude value of a light wave. Each parallel line in the diagram can be labeled alternately positive (+) and negative (−) according to amplitude values. The alternate pluses and minuses refer not only to the relative amplitude values at each wave front but also illustrate a simplified overhead view of a standing wave pattern. An overhead, or top, view of a standing wave pattern is a so-called grid with coordinates labeled alternately positive (+) and negative (−).

The grid diagram simply describes the interference between two sets of plane wave fronts of light. Wherever two maxima are added and wherever two minima are added, constructive interference takes place. Constructive interference is represented on the diagram by the intersection of two lines, both either labeled positive (+) or both labeled negative (−).

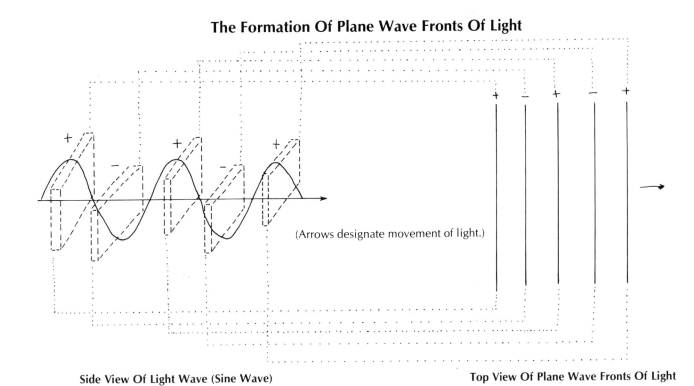

The Formation Of Plane Wave Fronts Of Light

(Arrows designate movement of light.)

Side View Of Light Wave (Sine Wave)   **Top View Of Plane Wave Fronts Of Light**

When a sine wave is intersected by planes perpendicular to the movement of light, so-called plane wave fronts of light are formed. These can be represented graphically by a set of parallel lines.

Wherever constructive interference takes place, there is also a corresponding addition of amplitude values. The amplitude addition, in this case, results in a greater total intensity of energy. As the two plane waves of light move through space, so do the points of intersection that represent points of constructive interference. Since the movement of a one-dimensional point through space produces a two-dimensional plane, the points of constructive interference become planes representing greater light intensity. These planes of greater intensity correspond exactly to the light fringes. The light fringes are always found to bisect the angle between the reference beam and the object beam.

Wherever a maximum and a minimum value of the amplitude are added together, destructive interference occurs. Destructive interference is represented on the diagram by the intersection of two lines, one labeled positive (+) and the other labeled negative (−). This amplitude addition results in zero electric intensity, which is, simply, darkness. It should be noted again that there is no cancellation of

The Interference Of Two Sets Of Plane Wave Fronts

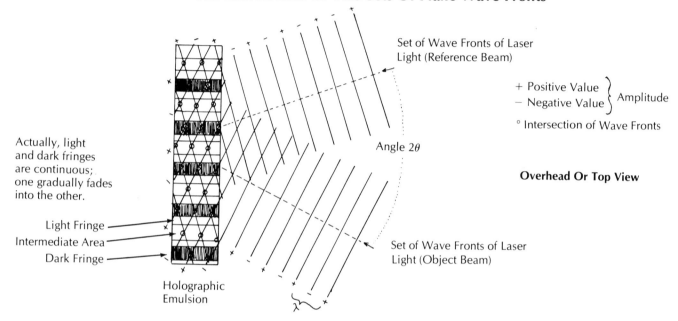

Set of Wave Fronts of Laser
Light (Reference Beam)

+ Positive Value
− Negative Value } Amplitude

° Intersection of Wave Fronts

Angle 2θ

Overhead Or Top View

Actually, light
and dark fringes
are continuous;
one gradually fades
into the other.

Light Fringe
Intermediate Area
Dark Fringe

Holographic
Emulsion

Set of Wave Fronts of Laser
Light (Object Beam)

λ

When two sets of plane wave fronts of laser light interfere, light and dark interference
fringes are formed throughout the holographic emulsion. Wave-front reconstruction
occurs when the reference beam illuminates the hologram.

electric charges and no cancellation of light waves. Zero intensity is simply a displacement of the light wave, which appears as absence of light to the human eye or, in a word, darkness.

When commercially available holographic plates or film are used, the light fringes are recorded when the extremely fine grains of silver halide are exposed to constructively interfering laser light. Energy from the laser light interacts with the electrons in the silver halide so that a latent image of the interference pattern is formed. The exposure to laser light and subsequent processing changes the silver-halide crystals of the latent image into the grains of silver that form the visible image. The grains of silver lie in planes that correspond exactly to the planes of greater light intensity produced by constructive interference. The planes of silver also mark the former position of the light fringes.

Likewise, the planes of zero intensity produced by destructive interference correspond to the dark fringes. The dark fringes indicate the absence of energy during exposure of the holographic plate to laser light. The unused silver-halide crystals, which are in the areas of destructive interference, are washed off when the holographic plate is fixed.

During the second step of the process, the hologram is exposed to laser light, which illuminates the holographic plate in the same way that the reference beam had during the recording. The laser light interacts directly with the planes of deposited silver that are found throughout the holographic emulsion. The planes of silver are very closely spaced and act like tiny mirrors in reflecting part of the reference beam. This reflected light forms the holographic image due to wave-front reconstruction. In this step, the hologram diffracts the reference beam into a reconstruction of the object-beam wave fronts. The reconstruction of the object-beam wave fronts appears as a three-dimensional holographic image. The holographic image is an exact rendition of the real object as far as visual appearance is concerned.

2-6 HOLOGRAPHY IN COMMUNICATION TERMS

In terms of communication, holography is a process of, first, information storage, and later, information retrieval. Information, in this sense, refers to the values of amplitude and the relative phase between coherent light waves. Coherent light waves mean laser light, of course.

Information storage takes place when an interference pattern of an object beam and a reference beam is recorded on holographic film. The record of wave-front interference is the hologram itself. The recorded interference pattern actually stores the amplitude and phase information from the two beams of light. This stored information consists of a three-dimensional holographic image of the object, complete in every detail.

Information retrieval takes place when the recorded interference pattern, the hologram, is again exposed to a coherent beam of light. When a beam of laser light illuminates the hologram, the hologram diffracts part of the light into an exact reproduction of the object beam. This reproduced object beam is the three-dimensional holographic image of the object. In this way, information in the form of a holographic image is stored in a recorded interference pattern and then is finally retrieved.

In terms of information storage and retrieval, both holography and radio communication have similar methods of handling information. Both techniques make use of some form of electromagnetic radiation. Holography, or more precisely optical holography, uses light waves that correspond to the optical or visible frequencies. Radio communication uses hertzian waves, which are the radio frequencies.

Both techniques of communication employ modulation and demodulation of electromagnetic energy. *Modulation*, in this sense, means superimposing different amplitude, frequency, or phase informa-

Communication: The Storage (Or Transmission) Of Information And The Retrieval (Or Reception) Of Information

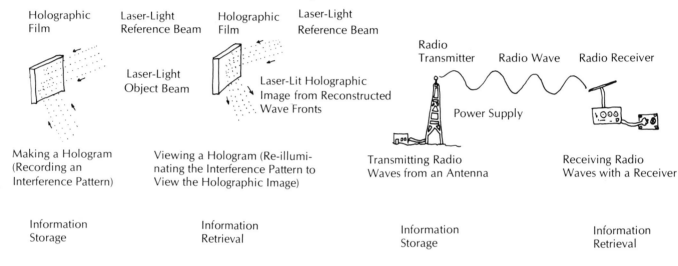

Optical Holography (At Visible Frequencies Of Electro-Magnetic Energy)

Radio Communication (At Radio Frequencies [RF] Of Electromagnetic Energy

tion on a carrier. A *carrier* refers to a steady and unvarying waveform that has the capacity to carry information. A carrier can be represented simply as an AC sine wave. *Demodulation* consists of retrieving the superimposed or added wave information from the carrier.

Most commonly, modulation of electromagnetic energy takes place at the radio frequencies. These include radio waves at medium through ultra-high frequencies ranging from 300 kilohertz to 3000 megahertz. A kilohertz, or kilocycle, is equivalent to 1000 cycles or 1000 vibra-

tions per second. A megahertz, or megacycle, is equivalent to 1,000,000 cycles or 1,000,000 vibrations per second.

The radio waves themselves are carriers that are modified or modulated by sound, or audio, information. Once they are modulated, the carrier waves are radiated into space. In communication terms, this corresponds to information storage or, more accurately, information transmission.

A radio antenna transmits the modulated carrier waves until they are picked up by a radio receiver tuned to a specific fre-

Radio Communication

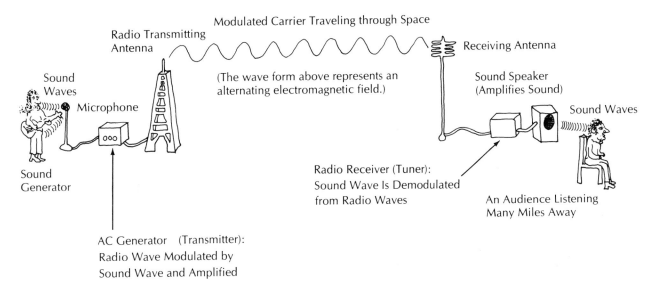

Modulated Carrier Traveling through Space

Radio Transmitting Antenna

(The wave form above represents an alternating electromagnetic field.)

Receiving Antenna

Sound Waves

Microphone

Sound Speaker (Amplifies Sound)

Sound Waves

Sound Generator

Radio Receiver (Tuner): Sound Wave Is Demodulated from Radio Waves

An Audience Listening Many Miles Away

AC Generator   (Transmitter): Radio Wave Modulated by Sound Wave and Amplified

At the radio station, sound waves enter a microphone. An alternating current generator produces electromagnetic energy at radio frequencies. This radio wave (the carrier wave) is modulated by the sound waves. A radio transmitting antenna radiates the modulated carrier until a radio receiver tuned to a specific frequency picks it up. The sound waves are demodulated by a detector, amplified by a sound amplifier, and emitted by a speaker. The human ear of a listener then receives the sound.

quency. The specific frequency refers to the particular frequency of radio waves emitted by the antenna. Demodulation within the radio receiver separates the sound information from the carrier wave. This demodulation produces sound and corresponds to information retrieval.

Modulation of electromagnetic waves occurs most commonly as *amplitude modulation* (abbreviated AM) and *frequency modulation* (abbreviated FM). AM and FM are well known as methods of radio communication.

Generally, in amplitude modulation,

information imposed on the carrier wave alters the amplitude values of the carrier itself. The amplitude values increase and decrease according to variations in the information signal. Modulation of the carrier causes two sideband frequencies with the same shape as the information signal to be produced. The two sideband frequencies are composed of one frequency that's higher and one that's lower than the carrier frequency. In a sense, the sideband frequencies enclose the carrier wave in what is called an *envelope of modulation*. The information signal itself is transmitted on

Amplitude Modulation (AM) At Radio Frequencies (RF) Of Electromagnetic Energy

Carrier Wave (Radio Frequency Signal)

Sound Wave (Information Signal):

**Carrier Wave Modulated By Sound
(Amplitude Modulated Wave)**

The carrier wave (above) is produced by an AC generator as oscillations of electromagnetic energy (single frequency and constant amplitude value). This type of wave (below) is composed of fluctuations of sound that modulate the carrier wave (different frequency of sound with varying amplitude values). The amplitude of the carrier is increased and decreased according to the change in amplitude of the sound wave. The carrier is contained in a so-called envelope of modulation (right).

the carrier frequency within this envelope of modulation.

In amplitude modulation at radio frequencies, a sound wave is the information signal that modulates the carrier wave. The carrier wave is a radio frequency signal produced by an AC generator. Fluctuations of sound modulate the carrier so that the envelope of modulation consists of sound waves with different amplitude values. The modulated wave envelope is broadcast or transmitted through the air. When the radio signal is received by a radio receiver, the carrier is demodulated and the sound waves are retrieved. Meanwhile, many miles away, an audience listens to the sounds.

The basic similarity between optical holography and radio communication is that each process uses a carrier that is modulated by an information signal. In radio communication, a radio wave acts as a carrier for information signals consisting of sound waves. In holography, a recorded interference pattern, which is the hologram itself, acts as the carrier. The carrier is modulated by amplitude and phase information that constitute the holographic image. In other words, the holographic image itself is the information that modulates the carrier at optical frequencies.

The recorded interference pattern, which acts as a carrier, is simply a holographically produced diffraction grating. A

Amplitude And Phase Modulation At Optical Frequencies Of Electromagnetic Energy (The Hologram As A Carrier)

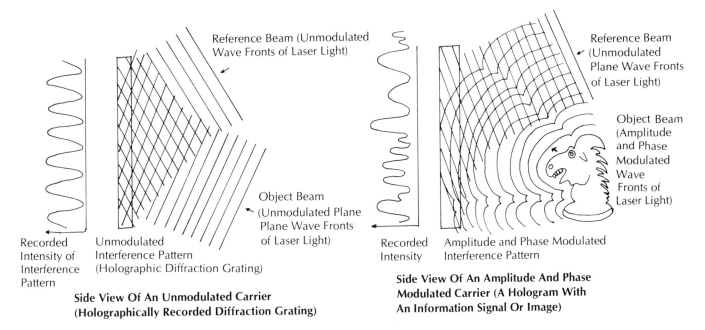

Reference Beam (Unmodulated Wave Fronts of Laser Light)

Object Beam (Unmodulated Plane Plane Wave Fronts of Laser Light)

Recorded Intensity of Interference Pattern

Unmodulated Interference Pattern (Holographic Diffraction Grating)

Side View Of An Unmodulated Carrier (Holographically Recorded Diffraction Grating)

Reference Beam (Unmodulated Plane Wave Fronts of Laser Light)

Object Beam (Amplitude and Phase Modulated Wave Fronts of Laser Light)

Recorded Intensity

Amplitude and Phase Modulated Interference Pattern

Side View Of An Amplitude And Phase Modulated Carrier (A Hologram With An Information Signal Or Image)

diffraction grating, which is a simple hologram, is made by recording the interference between two beams of laser light. Since the subject of such a hologram is unmodulated light—a single point of light—the hologram or, more accurately, the holographic grating, acts as an unmodulated carrier. What modulates the carrier is amplitude and phase information from light waves, which are reflected from an object. This information from an illuminated object consists of actual variations in the amplitude and phase of light. Modulation takes place when this object-beam information is recorded as actual variations in the structure of a simple diffraction grating.

Demodulation takes place when the grating, which is the hologram, is again exposed to laser light. If laser light of any visible wavelength illuminates the hologram from the same angle as in the recording, the hologram will diffract this light into a reconstruction of wave fronts. These reconstructed wave fronts make up the holographic image. The appearance of the holographic image means that demodulation of the signal from the carrier has occurred and that the information stored in the hologram has been retrieved.

In terms of holography, *amplitude modulation* of a carrier refers to variations in the amplitude of laser light waves that comprise the object beam. The variations in amplitude correspond to different values of intensity. The different intensity, or amplitude values, of object-beam light in effect modulates the recorded interference pattern that is the holographic grating itself. In other words, when the laser light waves with various amplitudes interfere with unmodulated light waves, the resulting hologram, or grating, is itself modulated. In communication terms, the carrier is modulated by the various amplitude values. In addition, the grating remains a time-withstanding record of its own amplitude modulation.

Phase modulation of a carrier consists of variations in phase between the object beam and the reference beam. The variations in phase correspond physically to variations in the spacing of the interference fringes. In other words, variations in phase correspond directly to variations in fringe width. The relative spacing, or width, of the fringes depends on the angle between the reference and the object beams according to the grating equation:

$$d \approx \frac{\lambda}{2\theta}$$

when d is the fringe width, λ is the wavelength of light, and 2θ is the angle between the reference beam and the object beam.

In regard to relative phase or fringe width, the grating equation means that when the angle between reference and object beams is large, the resulting fringe pattern is relatively fine. Inversely, as the angle between the two beams decreases, the spacing of the fringes becomes wider. The hologram itself is modulated by variations in phase and is also a relatively permanent record of the original relative phase between reference and object beams.

2-7 BASIC TYPES OF HOLOGRAMS

All holograms can be divided into two basic types: *transmission holograms* and *reflection holograms*. A transmission hologram is recorded when both the reference beam and the object beam are directed onto the same side of the holographic plate. A reflection hologram is recorded when the reference beam and the object beam approach the holographic plate from opposite sides.

The terms "transmission hologram" and "reflection hologram" both refer to the manner in which the hologram must be viewed. In the case of a transmission hologram, what happens is that laser light acting as a reference beam re-illuminates the hologram. The hologram diffracts part of this laser light into a wave-front reconstruction of the three-dimensional holographic image. The three-dimensional holographic image is transmitted through the hologram itself into the eye of the viewer. The transmission hologram is so named because the holographic image is seen by transmission. In other words, the light from the holographic image passes

Recording A Transmission Hologram   Recording A Reflection Hologram

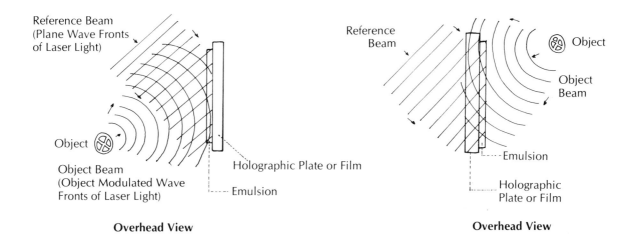

A transmission hologram is recorded when both reference and object beams approach the same side of the holographic plate or film (left). A reflection hologram is recorded when the reference and object beams are incident on opposite sides of the film (right).

through, or is transmitted by, the hologram.

In the case of a reflection hologram, what happens is that white light (not laser light) acts as a reference beam in re-illuminating the hologram. A holographic image is formed when the hologram diffracts part of the wave fronts of white light. This same holographic image then enters the eye of the viewer. Since the viewer is on the same side of the hologram as the incoming reference beam, the holographic image appears to be reflected directly from the hologram into the viewer's eye. Even though the light from the holographic image must pass through the hologram just as it does in a transmission hologram, the

apparent "reflection" of this light is supposed to account for the term "reflection hologram."

When a reflection hologram is re-illuminated by the same wavelength of laser light that was used in recording it, no wave-front reconstruction takes place. Therefore, no holographic image is formed. The reflection hologram is insensitive to the same wavelength of light because the recorded fringes have been distorted. The distortion occurs when the hologram is being photographically processed, specifically during fixing and drying. At that time, the holographic emulsion itself expands and contracts so that the fringes within the emulsion are distorted as well.

Viewing A Transmission Hologram   ## Viewing A Reflection Hologram

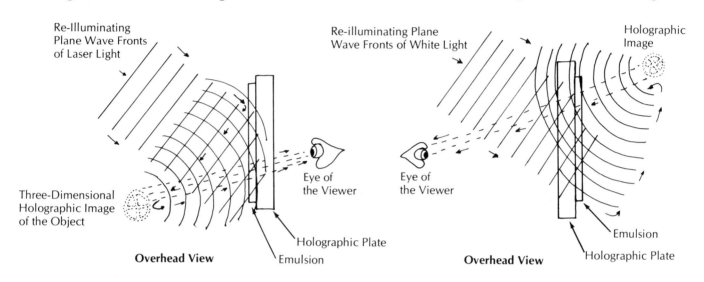

When a transmission hologram is re-illuminated by laser light, the hologram itself diffracts light so that a holographic image of the object is formed. The light from the holographic image then passes through the hologram into the eye of the viewer. Since this light is transmitted through the hologram, it is called a transmission hologram (left). When a reflection hologram is re-illuminated by white light, the hologram diffracts light into a reconstruction of the holographic image. Since the light appears to reflect from the holographic plate, this type of hologram is called a reflection hologram (right).

Because of the distortion, the fringes are incapable of letting the same wavelength of light pass through so that the hologram actually becomes insensitive to that wavelength.

The distortion of the spacing between the recorded interference fringes occurs when the hologram is being fixed. Fixing essentially removes undeveloped grains of silver halide from the gelatin emulsion. When the silver-halide crystals are removed, gaps are left between the deposited layers of silver. When the hologram is dried, the gelatin swelled by the water also dries rapidly so that the entire emulsion shrinks. The shrinkage of the emulsion causes the layers of silver inside the emulsion to shrink as well. Because of the shrinkage, there is a slight shift in wavelength sensitivity.

If the reflection hologram was recorded using 6328 angstrom red laser light, the shift in wavelength will be to the shorter green light around 5300 angstroms. In other words, the holographic image appears when the reflection hologram is illuminated by white light, which contains a wavelength of green light. The

The Formation Of Interference Fringes In Transmission And Reflection Holograms

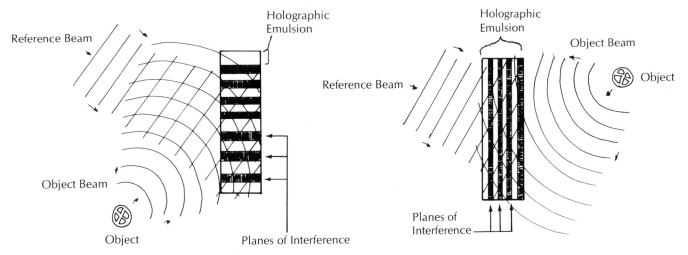

**Overhead View Of A Closeup Of Fringe
Formation In A Transmission Hologram**

**Overhead View Of A Closeup Of Fringe
Formation In A Reflection Hologram**

In a transmission hologram, the fringes lie in planes of interference perpendicular to the plane of the holographic plate. In a reflection hologram, the interference fringes lie in planes parallel to the plane of the plate or film.

wavelength of green light causes wavefront reconstruction to take place so that a green holographic image appears.

The shift in wavelength sensitivity in a reflection hologram can be predicted by the well-known *grating equation*. The grating equation states that the spacing between fringes is approximately equal to the wavelength of light divided by the angle between the reference beam and the object beam. Another way to state this relation is that the wavelength of light is approximately equal to the spacing between fringes multiplied by the angle between

the two beams. In symbols, this can be written:

$$\text{If } d \approx \frac{\lambda}{2\theta} \text{ then } \lambda \approx 2\theta \cdot d$$

when d is spacing between fringes, λ is the wavelength, and 2θ is the angle between the two beams. In other words, if the spacing between fringes shrinks or becomes smaller, the wavelength must be correspondingly shorter. This fits in exactly with the results, namely that a reflection hologram made with red laser light needs the shorter wavelength green light to produce a holographic image.

A reflection hologram, then, acts as a selective filter that allows only a certain wavelength to reconstruct the recorded wave fronts. Sources of white light used in re-illuminating a reflection hologram include penlights, slide-projector lamps, zirconium arc lamps, and natural sunlight. Because wave-front reconstruction occurs at frequencies of white light, reflection holograms are sometimes called *white-light-viewable holograms* or, simply, white-light holograms.

The fundamental difference between transmission holograms and reflection holograms is the relative position of the interference fringes recorded throughout the emulsion. The fringes are simply the layers of deposited silver that correspond to spaces within the emulsion where constructive interference took place. In the transmission hologram, the fringes lie in planes perpendicular to the plane of the holographic plate. In the reflection hologram, the fringes lie in planes that are parallel to the plane of the holographic plate. The planes of interference which are the layers of silver within a reflection hologram, are only about ½ wavelength apart. The fringes in a transmission hologram are usually spaced further apart. The holographic emulsion itself has an average width of 7 microns, or .007 millimeter thick.

Every transmission hologram and every reflection hologram is capable of yielding two holographic images. These are the primary image, sometimes called the virtual image, and the secondary image, sometimes called the real image. The secondary image is also called the *projected image* since this image seems to project from the holographic plate. The projected image occupies the space between the viewer's eye and the holographic plate, appearing to float in midair. The projected holographic image, however, is pseudoscopic, or simply inside out.

A hologram of the projected image can be made so that the characteristic inside-out image becomes orthoscopic, or right side up. Also, the projected image can be recorded so that it partially projects either in back or in front of the holographic plate or film. This type of hologram is called an *image-plane hologram,* or an image hologram.

When the reference beam approaches the film from the same side as the object beam (the object beam being the projected image itself), the hologram is called a transmission image-plane hologram. An image-plane reflection hologram is formed when the reference beam approaches the holographic plate from the opposite side as the light from the projected image.

Image-plane holograms can also be made by using a lens or a concave mirror to focus the light from a laser-illuminated object onto a holographic plate or film. Using one arrangement, a positive lens large enough to gather all the light from the object is placed between the object and the holographic film. The light scattered from the object passes through the lens and focuses either in front, in back, or directly in the plane of the holographic film. The lens, then, has the laser-lit object at one of its focal points and an image of the object at the other focal point. When a reference beam is added to the arrangement, a one-step image-plane hologram can be made.

In a similar arrangement, a concave mirror gathers the light from a laser-lit object and focuses it in the plane of the holographic film. With the addition of a reference beam that illuminates the entire holo-

Recording A One-Step Image-Plane Hologram (Using A Lens)

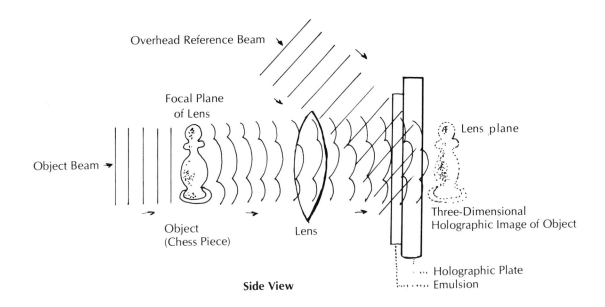

Overhead Reference Beam

Focal Plane
of Lens

Lens plane

Object Beam

Three-Dimensional
Holographic Image of Object

Object
(Chess Piece)

Lens

Side View

Holographic Plate
Emulsion

A one-step image-plane hologram can be made by illuminating an object with laser light so that the object light passes through a lens that has the object itself at one focal point and the image at the other focal point. With the addition of a reference beam, an interference pattern is recorded. Later, during wave-front reconstruction, the wave fronts of light from the image actually converge in the space in front of the holographic plate.

graphic film, a one-step image-plane holo-gram of another type is formed. Both types of one-step image-plane holograms use a collimated reference beam—laser light consisting of plane and parallel wave fronts. Also, both methods yield ortho-scopic images that are partly virtual, partly real. In other words, the holographic image from an image-plane hologram can be made to project in front of the holographic film, in back of the holographic film, or partly in back and partly in front.

Another commonly used method of making image-plane holograms is the two-step method. First, a regular transmis-sion hologram of a three-dimensional ob-ject or scene is made. This transmission hologram is called a *master-plate holo-gram* or, simply, a master hologram. In the second step of the process, the master hologram is re-illuminated by laser light so that the conjugate projected image is pro-duced. The light from the projected image itself is then used as the object beam in

Recording A One-Step Image-Plane Hologram (Using A Concave Mirror)

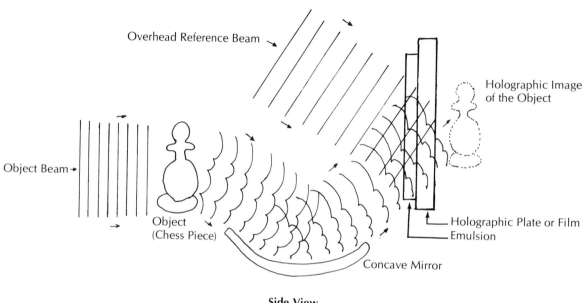

Side View

Another setup for recording a one-step image plane hologram can be made by illuminating an object with laser light so that the wave fronts of light from the object are reflected by a concave mirror. The mirror then focuses this object light at its focal plane. An interference pattern is recorded when the holographic plate or film is exposed to a reference beam as well as the mirror image of the object. During the wave-front reconstruction, the depth of the holographic image is somewhat limited, however, due to the effect of the concave mirror.

making an image-plane hologram. When a reference beam of unmodified laser light is added to light from the projected image, an image-plane hologram, which is actually a hologram of a hologram, is made. In this case, the image-plane hologram is also called a *copy hologram* since it is a copy of the master hologram.

Rainbow Holograms™ are a hybrid type of image-plane transmission holograms produced and developed by the School of Holography in San Francisco,

California. These image-plane holograms are also made with the two-step method, using two separate holographic camera setups.

The first step of the process consists of making a standard 4″ × 10″ transmission hologram. This hologram, which is called a master plate, is made by the aptly named Master Camera. The Master Camera is simply a transmission hologram setup—essentially lenses and mirrors mounted on a sand-table platform. The transmission

Recording A Two-Step Image-Plane Hologram (Making A Rainbow Hologram)
Step 1: Recording The Master Plane Hologram (A Transmission Hologram With Two Object Beams)

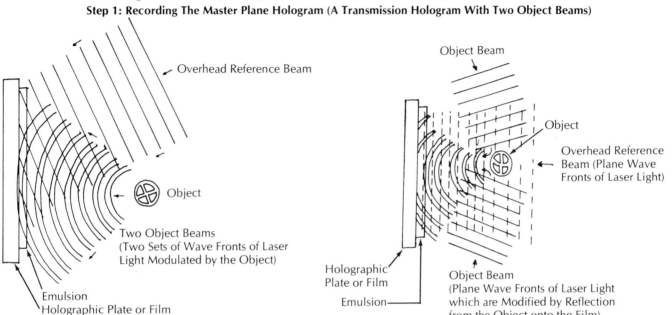

Side View

Overhead Reference Beam

Object

Two Object Beams
(Two Sets of Wave Fronts of Laser
Light Modulated by the Object)

Emulsion
Holographic Plate or Film

Overhead View

Object Beam

Object

Overhead Reference
Beam (Plane Wave
Fronts of Laser Light)

Holographic
Plate or Film

Emulsion

Object Beam
(Plane Wave Fronts of Laser Light
which are Modified by Reflection
from the Object onto the Film)

hologram made by using this setup is formed with an overhead reference beam. An overhead reference beam illuminates the entire holographic plate from above instead of from the side. Also, two object beams instead of only one are used in this setup. The two object beams are made by dividing a single object beam with a beam splitter. The resulting two beams of laser light can illuminate the object or scene from both sides, left and right, so that there is a much more even and uniform illumination.

In the second step of the process, the image-plane hologram is made in the so-called Rainbow Copy Camera. In this setup, a narrow horizontal band of laser light illuminates the master plate. Wavefront reconstruction occurs, thereby producing the projected image. The light from the projected image itself acts as the object beam for the copy hologram. When a collimated overhead reference beam is added to the light from the projected image, the

Making A Rainbow Hologram
Step 2: Recording The White-Light Viewable Copy (A Transmission Image-Plane Hologram)

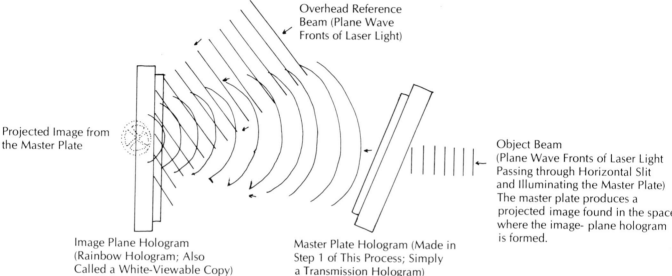

Overhead Reference Beam (Plane Wave Fronts of Laser Light)

Projected Image from the Master Plate

Object Beam (Plane Wave Fronts of Laser Light Passing through Horizontal Slit and Illuminating the Master Plate) The master plate produces a projected image found in the space where the image- plane hologram is formed.

Image Plane Hologram (Rainbow Hologram; Also Called a White-Viewable Copy)

Master Plate Hologram (Made in Step 1 of This Process; Simply a Transmission Hologram)

Rainbow Hologram™ can be recorded.

The holographic film holder in the Rainbow Copy Camera is adjustable. It can be moved either closer to, or farther away from, the master plate. The distance between the holographic film and the master plate determines the exact position of the holographic image in the Rainbow Hologram™. Depending on this distance, the holographic image can be made so that it projects in front of the holographic film, in back of it, or partly in back and partly in front.

The object beam of the Rainbow Hologram™ is formed by letting a beam of laser light pass through a cylindrical lens. The cylindrical lens spreads the point of laser light into a narrow horizontal band of laser light.

The purpose of the cylindrical lens is twofold. First, since only a horizontal band of laser light illuminates the master plate (instead of its entire surface), the object-beam light is more concentrated. Because of this, the resulting Rainbow Hologram™ is brighter and generally more efficient. Secondly, because of the slitlike object beam, the holographic image can be viewed with many sources of white light. The slitlike illumination during recording

Viewing A Rainbow Hologram

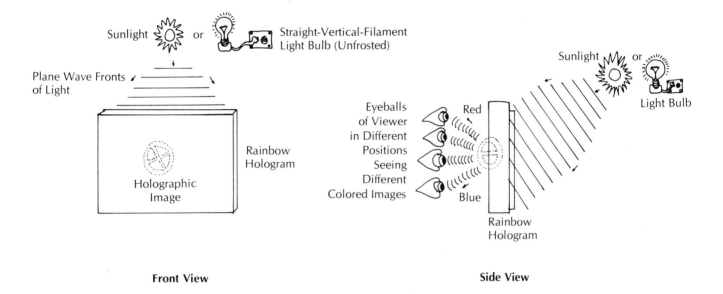

When white light in the form of sunlight or an unfrosted light bulb re-illuminates the hologram, wave-front reconstruction takes place. The viewer then sees the holographic image in each color of the rainbow, depending on the position of the eye in relation to the hologram.

allows the hologram to act as its own built-in filter or prism, which breaks up the white light during viewing. The Rainbow Hologram™ itself divides the white light into bands of color so that the holographic image appears in sequence in every color of the rainbow. The holographic image of a Rainbow Hologram™ appears in spectrally pure colors of red, orange, yellow, green, blue, indigo, and violet.

The holographic image of a Rainbow Hologram™ can be viewed with almost any undiffused source of white light, ranging from sunlight and candlelight to even full moonlight. The ideal source of light, though, is a bare, straight-filament light bulb—an ordinary unfrosted light bulb. Because of the high efficiency of a Rainbow Hologram™, sunglasses should be used in viewing the holographic image by sunlight. The hologram diffracts the sunlight so well that the resulting holographic image is so bright it's almost blinding. A light bulb, however, presents no such problem.

The only disadvantage of the cylindrical lens effect is that *vertical parallax* of the holographic image is slightly diminished. Vertical parallax is the ability of the observer to scan the holographic image

Making A Multiplex Hologram (Patent Pending)

Step 1: Making A 35 mm Motion-Picture Film Of A Subject Rotating Through 360°

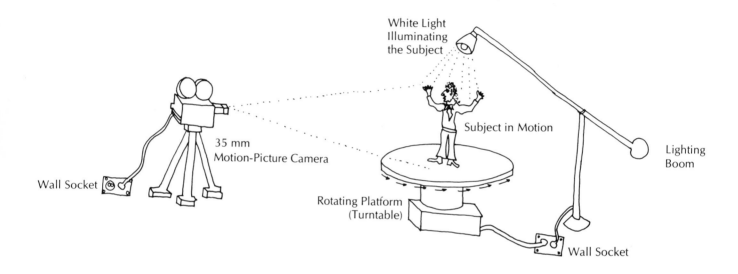

from top to bottom and to see different vertical perspectives of the holographic image. In other words, the ability of the observer to look over and under the holographic image is reduced. This disadvantage, though, is offset by the fact that the Rainbow Hologram™ can be seen with white light instead of laser light. This lowers cost tremendously and allows people without lasers to be able to look at holograms. Also, the cylindrical lens effect makes holograms much brighter than the holograms that are made and viewed with low-power lasers.

Another type of hologram produced and developed by the School of Holog-

raphy and Multiplex Co. of San Francisco, California, is the Multiplex Hologram (Patent Pending). The Multiplex Hologram is the result of a hybrid form of holography called *integral holography*. Integral holography is a technique that combines cinematography and holography to create three-dimensional holographic images of people and things in motion.

A Multiplex Hologram is actually composed of many narrow vertical slitlike holograms, side by side and parallel to one another. Each one of these narrow vertical holograms corresponds to a different view, or perspective, of the subject in motion. Because all the holographic images from

Making A Multiplex Hologram (Patent Pending)

Step 2: Making A Multiplex Hologram In The Multiplex Holographic Camera

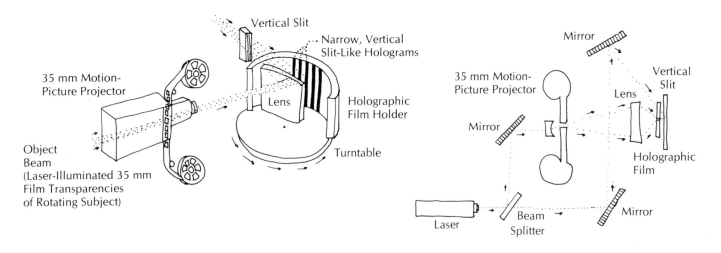

Perspective View Of Multiplex Camera

Side View Of Multiplex Camera

all the vertical holograms are integrated by the human eye into a single holographic image, the Multiplex Hologram is also called an integral hologram. The term "integral hologram" also refers to its composite nature since the holographic image from a Multiplex Hologram shows complete horizontal parallax. In other words, the holographic image of a subject in motion can literally be viewed from all sides. The holographic image moves according to the movement of the viewer's eyes so that every view of the subject—front, back, and sides—can be seen.

Making a Multiplex Hologram is also a two-step process. The first step con-

sists of making a 35mm motion-picture film of a subject rotating through 360°. The subject, which can be still or moving, is filmed on a rotating platform that turns around like a lazy Susan. The subject rotates completely around through 360° of movement. During the filming, usually three frames of movie film are shot for each degree of rotation. The resulting motion-picture film is a complete photographic record of a three-dimensional subject from each and every horizontal viewpoint. All views of the subject, from front to back, pass before the movie camera and are consequently photographed.

The second step of the process con-

Viewing A Multiplex Hologram (Patent Pending)

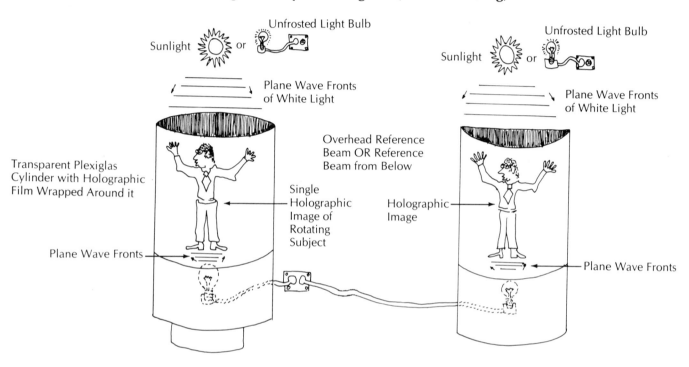

360° Multiplex Hologram Rotating-Cylinder Display   **120° Multiplex Hologram Display**

sists of making the Multiplex Hologram itself in the Multiplex Camera. The Multiplex Camera is a holographic camera that produces transmission holograms of any subject which can be filmed in a prescribed manner with an ordinary movie camera. Besides reproducing the movement of a three-dimensional subject, Multiplex Holograms can be viewed with white light instead of laser light.

In the Multiplex Camera, each individual frame from the 35mm motion-picture film is successively projected by laser light, just as a slide-film projector projects slides. Each laser-lit movie-film transparency constitutes the object beam for a single hologram. With the addition of an overhead reference beam, a series of successive holograms (one hologram per each motion-picture frame) is recorded on holographic film. In the case of a 360° Multiplex Hologram, about 1080 different holograms are recorded side by side.

Before the object beam reaches the holographic film, it passes through a specially made vertical slit. The vertical slit narrows down the image information from each movie frame so that a single vertical linelike hologram is recorded for each movie frame. The vertical slit also acts to integrate the many separate photographic images into a single holographic image.

This integrated holographic image portrays a three-dimensional subject moving through time. In a relative sense, the holographic image produced with a Multiplex Hologram is four-dimensional.

Besides passing through a vertical slit, the object-beam light also passes through a cylindrical lens just as in the Rainbow Hologram Camera setup. Similarly, a Multiplex Hologram can then be viewed either with sunlight or illumination by a straight-filament unfrosted light bulb. Also, like the image of a Rainbow Hologram™, the holographic image in a Multiplex Hologram appears in each of the colors of the rainbow.

Multiplex Holograms have been made of people (holographic portraits), people in action (for instance, a belly dancer, a couple drinking beer, a golf player swinging a golf club), three-dimensional models (architectural models of buildings, a molecular model of DNA, a topographic globe of our planet) as well as so-called art objects and sculptures. In other words, any subject accessible to cinematography is accessible to holography through the technique of integral holography. A Multiplex Hologram can be made of any subject that is filmed in a certain way and then subsequently processed in the Multiplex Camera.

Modes Of Oscillation Within A Laser Tube

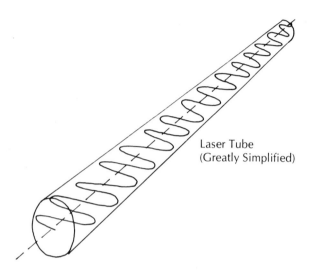

Laser Tube
(Greatly Simplified)

Transverse Mode
(Light vibrates from side to side or widthwise throughout the length of the laser tube.)

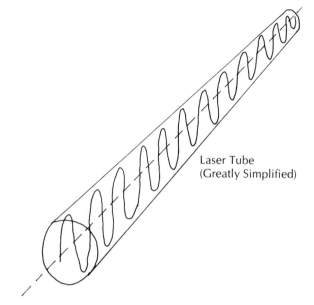

Laser Tube
(Greatly Simplified)

Longitudinal Mode
(Light vibrates lengthwise throughout the laser tube.)

2-8 REQUIREMENTS FOR MAKING A HOLOGRAM

There are several practical considerations that should be satisfied in order to record a hologram successfully. First, the temporal coherence and the spatial coherence of the laser light must be measured and tested in terms of the coherence length of the laser. Also, the relative intensity of the light from the object beam and the light from the reference beam must be measured. This measured relation between the intensity values of the two laser beams, called the light ratio, is used to determine the exposure time of the hologram. The holographic studio itself must be isolated in order to ensure the stability of the interference fringe pattern that is to be recorded. And finally, the general stability of the location for the holographic studio must be tested by setting up a simple interferometer.

The temporal coherence of laser light refers to the correlation between ordinary laser light and strictly monochromatic laser light. In general, ordinary laser light oscillates, or vibrates, in several modes and at slightly different frequencies. Monochromatic light usually means precisely single-frequency laser light. A *mode* refers to one of the ways in which laser light oscillates up and down and across the length of the laser tube. One mode is distinguished from another by the particular standing wave pattern the laser produces.

Most commonly, helium-neon lasers are adjusted for oscillation in a single transverse mode called the TEM$_{00}$ mode, an abbreviation for *transverse electromagnetic mode. Transverse* means that the oscillations occur widthwise throughout the length of the laser tube. The TEM$_{00}$ mode appears as a single dot or point of laser light.

Helium-neon lasers, which are most commonly used in holography, oscillate in several longitudinal modes. *Longitudinal* refers to oscillations of laser light that occur lengthwise throughout the laser tube. In other words, a laser that has several longitudinal modes produces laser light which is not, strictly speaking, a single frequency of light. Instead, this laser light actually consists of a band of frequencies, some slightly higher and some slightly lower than the central frequency, which is 6328 angstroms red. This type of laser light is called *multimode.* Multimode laser light is only an approximation of monochromatic laser light since there is more than one frequency of light.

Lasers that do not operate at a single frequency are nevertheless used in making holograms. The usual procedure is to divide a single laser beam into two beams of laser light with two different optical paths. An *optical path* is the route light travels through space in its movement from mirror to mirror and finally to the holographic plate. Because the two beams do not travel through the same space, the optical path of the object beam will be different than the optical path of the reference beam.

The degree of temporal coherence can be expressed in terms of the difference between the optical paths of the two beams of light. In other words, the coherence length of a laser can also be expressed by the difference in the two optical paths. The difference itself is the distance light travels in a certain time, that is, the wavelength.

In order to make a hologram using a non-single-frequency laser, the difference between the two optical paths must be less than the coherence length of the laser. The distance of each optical path is measured from the beam splitter, where the laser beam is initially divided, to the holographic plate, where the two beams finally interfere. When the difference between the lengths of the two optical paths is more than the coherence length of the laser, some of the wave fronts of object-beam light will be out of phase. Because some of the object beam will be different from the portions of the object which are out of range will not be recorded in the hologram, and the holographic image will be faded out in those areas.

Because of this fading effect, when a non-single-frequency laser is used to make a hologram the reference-beam distance and object-beam distance must be approximately equal. The distances of both optical paths must be measured with a tape measure, and the holographic components, lenses and mirrors, must be adjusted so that both distances are equal within a tolerance of about a couple of centimeters.

In a typical helium-neon laser with a laser tube 1 meter in length, the coherence length of the laser is also 1 meter. In comparison, a single-frequency laser with a laser tube 1 meter in length has a coherence length of 1 kilometer. Obviously, a single-frequency laser is much more desirable to use in making holograms. Measuring the reference-beam and object-beam paths becomes unnecessary. Moreover, the size of the subject is not restricted.

In order to make a single-frequency laser from a multimode laser, the laser tube must be coupled to another reflecting cavity called an *etalon*. An etalon consists of two partially reflecting plates of either quartz or glass, both plates parallel to each other. When laser light that oscillates in several longitudinal modes enters the etalon, the undesirable frequencies of laser light are in a sense filtered out. The output of a laser coupled to an etalon is practically single-frequency laser light. In holography, the use of etalons in conjunction with lasers is infrequent because of the prohibitive cost.

Although holograms can be made by using a single laser beam that acts as both reference beam and object beam, the usual procedure is to divide the laser beam into two separate beams of light. When the two beams of laser light interfere at the holographic plate, the relative intensity of each beam must be measured so that the exposure time for the hologram can be determined. The relative intensity of the reference beam in relation to the object beam can be measured by using a photometer, which is a light meter, or by using a converted voltmeter. The light meter measures both of the intensity values for the two beams separately. The intensity relationship between the two beams is called the *light ratio*.

In the case of a standard transmission hologram, the light ratio should be in a proportion of about 2:1 to 10:1, reference beam to object beam. The reference beam should be anywhere from two to ten times brighter than the object beam, since the reference beam in a sense acts as the carrier while the object beam is the information that modulates it. For best results in making a transmission hologram, a ratio of about 3:1 to 5:1 should be used. This light ratio will ensure a bright and clear hologram.

In the case of a reflection hologram, the optimum light-intensity ratio is 1:1 to 3:1, reference beam to object beam. After the ratio of light intensities has been measured, the exposure time for the hologram can be determined, since exposure time is directly dependent on the intensity of light. This procedure is similar to the light reading taken in photography. In photography, however, only a single value of light intensity is needed in most cases. Since the reference beam is brighter or more intense than the object beam, the value of the reference beam intensity is used with an exposure chart to determine a suitable exposure time. (See Appendix IV(C). The Light Meter.)

The general stability of the interference fringe pattern about to be recorded can be ensured by choosing a location for the holographic studio that is more or less free from vibrations caused by sound, heat, or any type of mechanical disturbances. These vibrations must be eliminated since any such disturbance will hinder photographing the interference fringes. Any movement of the fringes will effectively smear them, thereby displacing the energy that appears as a standing wave pattern. In other words, a hologram will not be recorded.

The stability needed to record a hologram is about ¼ wavelength tolerance. This means that the stability between all holographic components (beam splitter, mirrors, lenses, and holographic plate holder) must be such that the interference fringes do not shift more than ¼ wavelength during the exposure. With these considerations in mind, the ideal location for a holographic studio is a space filled with silence and removed from all sources of mechanical and sound vibrations. Practically, the holographic studio should not be anywhere near streetcar tracks, an airport, a rock-and-roll band, a train station, and the like.

To further isolate the holographic camera from random vibrations, it is common practice to use a so-called isolation table. Isolation tables range from granite slabs suspended on partially inflated inner tubes to cast-iron optical benches to sandboxes raised to table height. The holographic components, that is, the laser, mounted lenses, mirrors, beam splitter, are housed on these isolation tables. A primary consideration in building an isolation table is to use materials that are available locally since those materials are usually the least expensive. The sandbox-type isolation table is probably the most popular and most extensively used since sand is readily available and relatively inexpensive.

The general stability of the location that is chosen for the isolation table can be determined by performing a simple test called an *interferometer*. An interferometer literally means measuring interference. An interferometer can be set up to test the stability of a particular set of holographic components, an isolation table, the floor on which an isolation table is to be built, or the general location of the building that houses the holographic studio.

The interferometer is basically a setup for the observation of interference

Using An Etalon To Produce Single-Frequency Laser Light

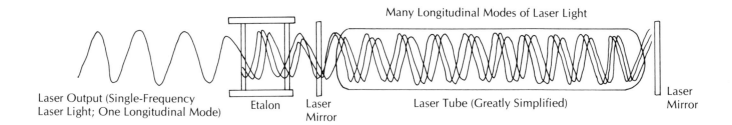

Laser Output (Single-Frequency
Laser Light; One Longitudinal Mode)

Etalon   Laser
          Mirror

Many Longitudinal Modes of Laser Light

Laser Tube (Greatly Simplified)

Laser
Mirror

fringes produced by the interference of a point source of object light and a point source of reference light. The interference patterns of both beams of light are superimposed one on the other. The superimposition creates a single Fresnel zone pattern of circular, concentric, alternately light and dark fringes. This Fresnel zone pattern can be observed as an indication of the stability of the vicinity. Even touching the holographic components or the isolation table itself causes the fringe pattern to implode or explode. In other words, the light fringes move into the space where the dark fringes are, and vice versa. When this happens during exposure of a holographic plate or film, no hologram is formed because the fringes are smeared.

The interferometer is such a sensitive test that even breathing on the components will disrupt the interference fringes. Hot breath is not an asset in making holograms. When an interferometer has been successfully performed, the formation of a hologram can be assumed.

CHAPTER 3
BASIC HOLOGRAPHIC TECHNIQUE

3-1 INTERFEROMETER SETUP

A simple interferometer can be used to test the general stability of the isolation table or, more importantly, the proposed site for an isolation table. In fact, an interferometer test should always be performed at the proposed site of a holographic camera prior to any actual construction of the isolation table itself. This will ensure that the location is indeed interferometrically stable enough to produce holograms. Actual holograms need not be made at this stage since the stability can be tested by direct observation of the interference fringes. A listing of the holographic components needed to assemble an interferometer and the actual procedure in step-by-step instructions follows. Although the instructions refer to a sandbox-based holographic camera system, they are equally applicable for any other type of isolation table or other location.

Holographic components used in an interferometer (not including the laser):

3 mirrors (front-surfaced and colli-mated)
1 spreading lens bi-concave, focal length: approx. -100 cm
or
1 microscope objective ($60\times$)
1 beam splitter (50/50)
1 plate holder
1 cardboard screen (piece of cardboard cut to dimensions of plate holder)

Procedure for assembling an interferometer:

1. Place a mirror in the corner of the table to direct the laser beam onto the table. Adjust the mirror so that the beam approximately bisects the plane of the table into triangles. Using simple geometry in

Interferometer Setup

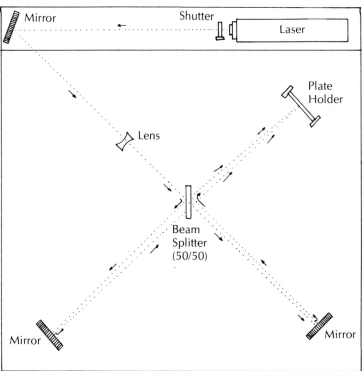

Overhead View Of Isolation Table

(Arrows signify the direction of the movement of laser light; dots designate the laser-light path.)

designing the positions of the holographic components makes it easy to adjust distances later.

2. Place another mirror in the corner of the table opposite the first mirror. Align it so that the reflection from the two mirrors are exactly superimposed. In other words, adjust the mirrors so that only one spot of laser light appears on each mirror. When using a sandbox isolation table, make sure the components stand vertically in the sand (perpendicular to the plane of the surface of the sand),

otherwise the components have a tendency to drift out of alignment.

3. Place the beam splitter in the middle of the sandbox between the two mirrors so that the laser beam enters the beam splitter at approximately a 45° angle. Then, half of the beam is transmitted and half reflected to the side, so an angle of 90° is formed between the incoming laser beam and the beam reflected by the beam splitter.

4. Place another mirror in the corner of the table to receive the beam reflected by the

beam splitter. This mirror should be approximately the same distance from the beam splitter as the mirror that receives the light transmitted through the beam splitter.

5. Adjust the mirrors so that all the spots of light are superimposed (on top of each other).

6. Position the plate holder so that the distance from each of the mirrors to the beam splitter is equal to the distance from beam splitter to plate holder. Insert the white cardboard in the plate holder and adjust the mirrors so that all beam spots are superimposed.

7. Place a spreading lens or microscope objective lens between the first mirror and the beam splitter so that a set of Fresnel zones or interference fringes appears on the white card in the plate holder.

If a set of circular concentric alternating black and red (in the case of a He-Ne laser) rings does not appear on the white card, remove the lens and check to see if all the spots of light are centered in the middle of the card on top of each other. Check the distance; the distance from the beam splitter to a corner mirror then back to the beam splitter and to the plate holder should equal the distance from the beam splitter to the other corner mirror then through the beam splitter to the plate holder.

The Fresnel zones that appear on the white card are due to the interference of two beams of laser light which are separated by an angle that is effectively zero. Because of this, the size of the fringes is greatly magnified so that they can be observed with ease. Observing the fringes is a simple test of the stability of the setup, the isolation table, and the isolation in general. The extreme sensitivity of the fringes can be observed by lightly nudging the isolation table, stamping on the floor, and even breathing on the components, since any kind of gross vibration, either thermal or mechanical, will disturb the fringes and cause them to blur. When this happens during exposure to a holographic plate, no hologram is recorded, since recording the interference fringes, or standing wave pattern, is essentially making a hologram.

To translate the intensity of the interference fringes (visual information) into audio information, a photosensitive cell can be used in the place of the white card in the plate holder. Coupled with an audio amplifier and a loudspeaker, the photocell allows the visual signal to be heard. Pulses of sound (signaling a shift in the standing wave pattern) seem to be more expressive of the actual sensitivity or delicacy of the fringes and more impressive than scanning the pattern with the human eye.

3-2 TRANSMISSION HOLOGRAM SETUP

3-2.1 RECORDING A TRANSMISSION HOLOGRAM

A standard transmission hologram setup can be made by dividing the light of a single laser beam into two parts: the *reference beam* (laser light that is directed onto a holographic plate) and the *object beam* (laser light that illuminates the object so that the light from the object is reflected onto the holographic plate). In a transmission hologram, both reference and object beams are directed onto the same side of the holographic plate or film. When more than one object beam is desired (to illuminate the object more fully and to reduce the shadow or darkness content of the hologram), another beam splitter can be used to form a double-object-beam transmission hologram. A listing of the holographic components needed to assemble a single-object beam as well as a double-object-beam transmission hologram setup and the procedure in step-by-step instructions follow.

Holographic components used in a single-object-beam transmission hologram setup (not including the laser):

    3 mirrors (front surfaced and collimated) 2″×3″
    4 spreading lenses (concave)
                or
    2 microscope objectives (60×)
    1 beam splitter (90/10)
    1 diffusion screen (piece of diffusing plastic or groundglass screen)
    1 plate holder
    1 cardboard screen
    1 box of holographic plates: Agfa-Gevaert 10E75 or 8E75, suitable for He-Ne lasers
    1 light meter
    1 tape measure

optional:
    1 pinhole or spatial filter assembly

To assemble a double-object-beam transmission hologram setup, add:

    1 beam splitter (50/50)
    1 mirror
    2 lenses
    1 diffusion screen

Procedure for assembling a transmission hologram setup:

1. Place a mirror in the corner of the table. This mirror can be mounted more permanently by clamps or glue.
2. Place the plate holder on the table on the side opposite the laser and approximately parallel to the position of the laser.
3. Place the object several inches in front of the plate holder. The object or objects (in the case of a scene) can be placed on the sand or on a small platform covered with dark cloth (for greater contrast). The height of the platform (simply a piece of wood with a piece of tubing or pipe attached to the bottom) can be adjusted to the height of the plate holder so that the height of the laser beam, the plate holder, and the object are approximately the same. This affords a degree of precision on the shifting sands of the isolation table.
4. Place the object-beam mirror in the corner of the table opposite the laser-transfer mirror so that the point of laser light illuminates the object.
5. Place one or two lenses in the path of the laser beam to spread out the laser light so that the object is fully illuminated. A diffuser can be placed in front of the lens to eliminate hot spots (glare) on the laser-lit object and to ensure an even illumination. In effect, the diffusion screen makes the coherent laser light illuminating the object incoherent. This causes each point of light which radiates from the laser-lit object to emit spherical wave fronts that, unlike plane wave fronts, man-

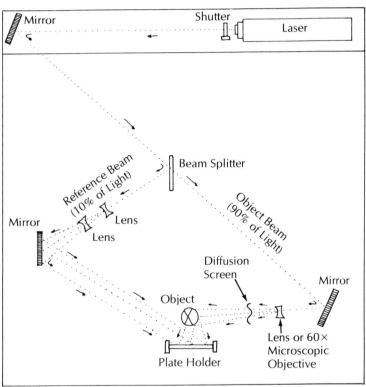

Transmission Hologram Setup (One Object Beam)

Mirror · · · Shutter · · · Laser

Beam Splitter

Reference Beam (10% of Light) · Lens · Lens

Object Beam (90% of Light)

Mirror

Diffusion Screen

Mirror

Object

Lens or 60× Microscopic Objective

Plate Holder

Overhead View Of Isolation Table . . . . Laser Light Path
→ → → Direction of Laser Light Movement

age to illuminate the entire holographic plate.

If two object beams are desired, proceed to step 5a; if not, proceed directly to step 6.

5a. Place a 50/50 beam splitter approximately in the middle of the table, to divide the object beam into two parts.

5b. Place a mirror in the corner of the table approximately opposite the other object-beam mirror and direct this beam onto the object.

5c. Place one or two lenses in the path of this object beam (and a diffuser) and

light the object to satisfaction. Looking through the plate holder as if through a window, adjust the laser illumination. What the holographer sees through the plate holder is what the holographic plate will see and record, so that any changes in the object illumination will be recorded as the interference pattern emanating from a laser-lit object.

6. Place a 90/10 beam splitter in the path of the laser light; 90 per cent of the light will pass through and will be

Transmission Hologram Setup (Two Object Beams)

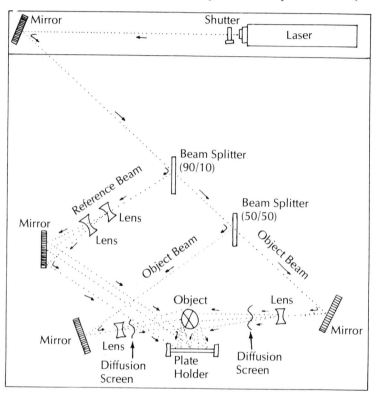

Overhead View Of Isolation Table

. . . . Laser Light Path

——————→ Direction of Laser Light Movement

used as the object beam, 10 per cent is used as the reference beam. The position of this beam splitter should be between the laser-transfer mirror and the object-beam mirror, or when two object beams are used, between the laser-transfer mirror and the 50/50 beam splitter (about halfway between).

7. Place a reference-beam mirror in the path of the laser light and direct this light onto the cardboard screen in the plate holder.

8. Measure the reference- and object-beam path lengths and adjust these distances so they are equal to within ½ inch. The object-beam distance (measured from 90/10 beam splitter to object-beam mirror to the center of the object to the center of the cardboard screen) should equal the reference-beam distance (measured from the 90/10 beam splitter to the reference-beam mirror to the cardboard screen in the plate holder). If the distances are not equal, move the reference-

beam mirror and adjust the reference-beam distance to the object-beam distance. When two object beams are used, both object-beam distances (measured from 90/10 beam splitter to the plate holder) must be equal to themselves as well as the reference-beam path (measured from 90/10 beam splitter to the plate holder). If all the optical path distances are not equal, laser light arriving at the holographic plate during exposure will be out of phase resulting in no hologram or a partial hologram.

9. Place one or two spreading lenses or a 60× microscope objective in the path of the reference beam and expand the beam so that it covers the entire cardboard screen. It may be helpful to block the object beam temporarily by placing a piece of cardboard in the object-beam path immediately after the 90/10 beam splitter. This facilitates seeing the reference beam as it is without the object light.

10. Measure the ratio of reference-beam intensity to object-beam intensity with a light meter (either a Triplet V-O-M Meter or an S. & M. Darkroom Meter have been used satisfactorily). The ratio of reference to object beam should be anywhere from 2:1 to 10:1 for a transmission hologram (optimum 3:1 to 5:1). If the ratio is somewhat off the desired value, adjust the intensity of the light by moving the lenses either closer together or farther apart. Lenses that are closer together tend to spread the light less than lenses that are farther apart. Also, lenses close together tend to produce an intensity greater than if they were placed at a distance from each other.

Addition of any lenses reduces the intensity of the light (about 10 per cent loss of light intensity per lens). When the ratio has been adjusted to satisfaction, an exposure chart can be consulted for the appropriate exposure time, using the value of the reference-beam intensity to determine the length of exposure. (See Appendix IV (C). The Light Meter.)

The procedure for using a light meter follows:

10a. Turn off all the lights except the laser.
10b. Block the object beam by placing a piece of black cardboard in the object-beam path between the 90/10 beam splitter and the object-beam mirror.
10c. Remove the cardboard screen from the plate holder and place the photocell of the light meter in the middle of the plate holder, tilting the photocell in the direction of the reference beam. It is even easier to cut a hole in the center of the cardboard screen large enough for the photocell to poke through. A center hole also facilitates centering the reference beam on the cardboard.
10d. Read the photometer starting with the scale for lowest light intensity. If there is no reading, change the scale until there is a registration of light on the meter.
10e. Remove the cardboard blocking the object beam and block the reference beam by placing the cardboard between the beam splitter and the reference-beam mirror.

10f. Read the photometer and correlate this reading to the previous reading, thereby obtaining a ratio of reference to object beam intensities of light.

10g. Remove the cardboard blocking the reference beam.

11. Replacing the cardboard in the plate holder, take note of the pattern resulting from both reference and object beams. If there are random interference patterns on the cardboard, it is probably due to dust or dirt on the lenses or mirrors. Lightly clean all holographic components with lens tissue. Make sure all components are firmly placed in the sand; pile sand around the bottom of each component to minimize drift during exposure. Leave the isolation table area for 10 minutes (*presettling time*) to ensure that none of the components is drifting.

12. Check the pattern on the cardboard and correlate it to the earlier pattern. If the pattern appears to be stationary, turn off the light.

13. Check the laser shutter to make sure it is working properly by turning it on and off. (See Appendix III(B). Construction of Holographic Studio Components, for reference to the laser shutter.)

14. Remove the cardboard from the plate holder and turn off the laser shutter.

15. Load the holographic plate or film in the plate holder, with the emulsion side toward the object. The emulsion side of the plate can be determined by placing a corner of the plate between the lips (lightly kissing the plate). One side of the plate (the emulsion side) will stick to one lip.

16. Leave the isolation table area for 10 minutes to allow it to settle and to allow any mechanical, thermal, or other vibrations enough time to disappear. This is called the *settling time*.

17. Expose the holographic plate to the laser light by turning the shutter on for the time determined according to the exposure chart. If you intend to bleach, you will need a somewhat longer exposure. (See Appendix IV(C) for further reference to exposure.)

18. Replace the exposed plate in a light-proof box. The plate is ready for processing—developing, stopping, fixing, washing, bleaching, and drying—like any ordinary black-and-white photograph with a silver-halide base emulsion.

3-2.2 PROCESSING A TRANSMISSION HOLOGRAM

Processing a transmission hologram recorded on a silver-halide base film (Agfa-Gevaert 10E75 holographic plates or film) consists of developing, stopping, and fixing the hologram in the same way that a standard black-and-white photograph is processed. A listing of darkroom chemicals and other supplies as well as the actual procedure of processing a transmission hologram follows.

Darkroom chemicals and supplies used in processing a transmission hologram:

1 can fine-grain developer (Kodak D-19, Kodak Microdol-X, Kodak Dektol, Kodak D-11, or Acufine)
1 can Kodak Indicator Stop Bath
1 can Kodak Rapid Fixer
1 stirring rod

6 processing trays or plastic buckets
1 bottle Kodak Photo-Flo
1 squeegee
1 hair dryer (for drying)
2 holographic plate holders

optional:

1 timer or watch

Procedure for processing a transmission hologram:

1. Before exposing the holographic plate, prepare all chemicals necessary for processing the hologram. Mix up the chemicals in gallon jugs to store, then fill a processing tray with each chemical, one each for developer, stop, fix, and Photo-Flo, and two for water cleansing baths—one for washing after the stop and one for washing after the fixer. It is also advisable to mix up a solution of bleach for the holographic plate since bleaching the plate improves the diffraction efficiency and makes the hologram brighter and easier to see. (Recipes for preparing bleaching solutions suitable for transmission holograms recorded on silver-halide base film can be found in Appendix II(B).)

2. Enter the darkroom with the exposed holographic plate still protected by the lighttight box. Make sure the overhead lights are turned off and the safety light is on (a green light, for example, Kodak Wratten Series 0A or Series 3, is particularly effective as a safety light since the holographic film is sensitive mostly to red).

3. Transfer the plate to a plate holder if buckets are used for processing chemical solutions. When developing trays are used instead, note which side of the plate is the emulsion side and place the plate in the developing tray emulsion side up. Agitate thoroughly to ensure a short developing time. Check every 30 seconds and stop development when the plate is a light to medium gray. If the plate doesn't turn gray within five minutes, the exposure probably was not long enough, or there might have been a piece of cardboard in the path of one of the beams. If the plate turns dark in less than one minute, the exposure was probably too long. The optimum time for a transmission hologram in the developer is about two to five minutes.

4. Transfer the plate from the developer to the stop-bath solution and agitate the plate for about 30 seconds. The stop bath stops the developing action of the developer.

5. Transfer the plate from the stop-bath solution to a tray or bucket of water to wash any developer or stop solution off the plate and prevent contamination of chemical solutions. Agitate the plate in the wash for about 30 seconds.

6. Transfer the plate from the first wash to the fixer solution. Agitate the plate for a couple of minutes. Leave the plate in the fix for about ten minutes so that all the undeveloped silver-halide crystals are removed from the plate.

7. Transfer the plate from the fix into another water-filled container, the second wash. Leave the plate in this wash for about ten minutes.

8. To check if there is any holographic image at all on the plate, examine the plate by looking through it at a white light bulb. Tilt the plate until a rainbow smear appears. This rainbow smear is the holographic image in all colors of

the rainbow, since white light is used to reconstruct the recorded wave fronts. Using a laser to view the hologram in a sense filters out only the red wavelength so that the image appears in red and not as a rainbow of colors. If there is no rainbow smear, it is a sure indication that no hologram was recorded or developed.

9. Transfer the plate from the second wash to the Photo-Flo and leave it there for about five to ten minutes. The Photo-Flo simply helps wash all the old chemicals off the plate.

10. Wipe the plate with a photo sponge, or squeegee, and dry it. It can be left to air-dry or, faster yet, the plate can be dried by using a hair dryer. The plate is ready to be examined with laser light.

3-2.3 VIEWING A TRANSMISSION HOLOGRAM

The processed transmission hologram can be viewed either prior to or after bleaching. Before bleaching, the holographic plate can be thoroughly examined to determine what happened during the setup, exposure, and processing. The procedure for viewing a transmission hologram follows.

1. Place the holographic plate, emulsion side toward the object, back in the plate holder.

2. Block the object beam by placing a piece of cardboard in its path. For a brighter reference beam, the beam splitter can be replaced by a mirror (when the beam splitter is 90/10) or tuned to maximum reference beam brightness (when the beam splitter is a variable one). The holographic image due to wave-front reconstruction can be clearly seen. When the reference and object beams are alternately blocked, the holographic image can be compared to or superimposed on the object. At this point (if the hologram was properly recorded), it is difficult to distinguish visually between the holographic image and the three-dimensional object itself.

3-3 SINGLE-BEAM TRANSMISSION HOLOGRAM SETUP

The single-beam transmission hologram setup can be made without using a beam splitter to divide the laser light. Instead, a single laser beam acts as both reference and object beam. This is done by expanding a single laser beam with a spreading lens or microscope objective. In this way, an area or space on the isolation table surface where objects are placed as well as the plate holder are simultaneously illuminated by laser light. When objects placed in the path of the laser light are situated below the plate holder (so as not to cast a shadow on the plate), laser light illuminates the objects, which in turn reflect light onto the holographic plate (the object beam). The laser light that passes over the objects directly illuminates the plate (the reference beam). This type of hologram is uncomplicated to set up and desirable because of the long depth of field

Single-Beam Transmission Hologram Setup

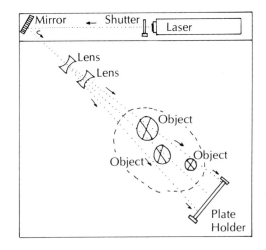

Overhead View Of Isolation Table

. . . . Laser Light Path
———→ Direction of Laser Light Movement
- - - - - - - - - - Sand Level (Dugout Area)

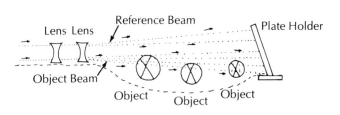

Side View Of Isolation Table

(up to several feet). A listing of the required holographic components and step-by-step instructions follows.

Holographic components used in a single-beam transmission hologram setup (not including the laser):

    1 mirror
    2 spreading lenses
           or
    1 microscope objective (60×)
    1 plate holder
    1 cardboard screen
    1 light meter

    1 box holographic plates: Agfa-Gevaert 10E75
optional:
    1 pinhole or spatial filter assembly

Procedure for assembling a single-beam transmission hologram setup:

1. Place the laser-transfer mirror in the corner of the table to direct the laser beam onto a sandbox-type isolation table.

2. Place the plate holder on the table in the corner opposite the laser-transfer mirror. Tip the plate holder to the 30° position so the plate faces the objects.

3. Dig out the sand in an area approximately 1 foot or 1 meter by several inches directly in front of the plate holder.

4. Place the objects in this dug-out area so that they are below the plate holder. In this way, no shadows from the illuminated objects are cast on the plate.

5. Place one or two lenses in the path of the laser beam to spread out the light so that all of the objects are illuminated. The top portion of the beam illuminates the cardboard screen in the plate holder (reference beam) while the bottom portion illuminates the objects in the sand pit (object beam). A 60× microscope objective can also be used instead of the lenses. However, diffusion screens should not be used in making a single-beam transmission hologram since they would scramble or randomize the reference beam so that the wave fronts from the object recorded in the hologram could not be retrieved.

6. Since no measurement of reference- and object-beam path distances is necessary (they are both essentially the same beam and approximately equal path lengths), measure the ratio of reference to object beam intensities by using a light meter. The reference-beam intensity can be measured by blocking out the object beam, that is, the lower half of the plate. This is done by covering the lower half of the plate holder with a piece of cardboard and placing the photosensitive cell of the light meter perpendicular to the plane of the plate and centered. The intensity of the object beam can likewise be measured by blocking out the reference beam, covering the upper half of the plate holder with the cardboard and placing the photosensitive probe in the center of the uncovered space. To adjust the intensity of the single beam so that either the reference or object beam is more intense, take out the lenses and adjust the laser-transfer mirror by moving it vertically until the more desired distribution of intensity is reached. This ratio of reference to object beam should be 2:1 to 10:1.

7. Leave the isolation table area for the standard presettling period to ensure that none of the components is drifting (about five minutes).

8. Check the laser shutter by turning it on and off. If the shutter is manually controlled, block the laser beam with a cardboard screen; if automatic, turn off the laser shutter.

9. Load the holographic plate or film (with the emulsion side toward the object) in the plate holder.

10. Leave the isolation table area for five to ten minutes to allow the table to settle.

11. Expose the holographic plate to the laser light by turning the shutter on for the time determined by the light ratio and exposure chart.

12. Replace the exposed plate in a light-proof box and process it according to the directions for processing a standard transmission hologram.

13. To view the hologram, replace it in the plate holder on the isolation table. The holographic image may be initially difficult to see since both object and reference beams are the same beam of laser light. In a sense, the viewer must look into the laser light to see the reconstructed holographic image.

Reflection Hologram Setup (One Object Beam)

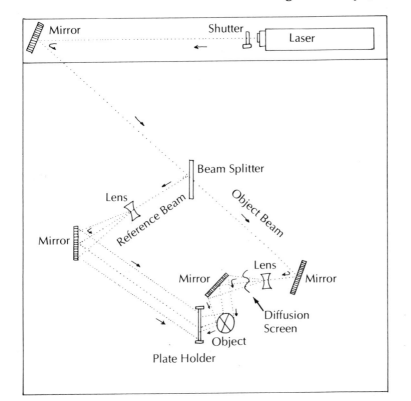

Overhead View Of Isolation Table

. . . . Laser Light Path
———→ Direction of Laser Light Movement

3-4 REFLECTION HOLOGRAM SETUP

3-4.1 RECORDING A REFLECTION HOLOGRAM

A standard reflection hologram setup can be made by dividing a single laser beam into an object beam and a reference beam. Unlike a transmission hologram setup (in which both beams approach the holographic plate from the same side), a reflection hologram is made by directing the reference beam onto one side of the plate and the object beam onto the other side. In this way, the recorded interference fringes are formed parallel to the plane of the plate so that in the reconstruction step of the process, the holographic image is viewed by reflection. The light source, acting as the reference beam, illuminates the recorded interference fringes (the hologram), which then reflect light in the form of the holographic image into the eye of the viewer.

When more than one object beam is desired, in order to illuminate the object from two sides instead of one, another beam splitter can be added to the setup, resulting in a double-object-beam reflection hologram. A listing of the holographic components necessary to the assembly of single-object beam and double-object-beam reflection hologram setups as well as the actual procedure follows.

Holographic components used in a single-object-beam reflection hologram setup (not including the laser):

> 3 mirrors
> 4 spreading lenses
> > or
> 2 microscope objectives (60×)
> 1 beam splitter (50/50)
> 1 diffusion screen
> 1 plate holder
> 1 cardboard screen
> 1 box of holographic plates: Agfa-Gevaert 8E75 NAH (no antihalation backing)
> 1 light meter
> 1 tape measure

optional:
> 2 pinholes or spatial filter assemblies

To assemble a double-object-beam reflection hologram setup, add:
> 1 beam splitter (50/50)
> 1 mirror
> 2 spreading lenses
> > or
> 1 microscope objective (60×)
> 1 diffusion screen

Procedure for assembling a reflection hologram setup:

1. Place the laser transfer mirror in the corner of the table near the laser to direct the laser beam onto the table.
2. Place the plate holder on the table on the side opposite the laser and approximately perpendicular to the position of the laser. It is helpful when the plate holder is centered in this position with respect to the isolation table.
3. Place the object several (2 to 3) inches away from the plate holder.
4. Place the object-beam mirror in the corner of the table opposite the laser-transfer mirror.
5. Place another object-beam mirror near the object so that laser light can be transferred from the mirror in the corner of the table to the second mirror. This dogleg in the object-beam light path enables the object to be illuminated from the front and can be adjusted to correspond to the reference-beam path distance. Center the point of laser light in the middle of the cardboard screen that is resting in the plate holder.
6. Place one or two lenses in the path of the laser beam (between the two object-beam mirrors). When the object has been illuminated to satisfaction, place a diffusion screen between the object-beam mirror and the object, if even illumination with no hot spots on the object is desired.

If two object beams are desired, proceed to step 6a; if not, proceed directly to step 7.

6a. Place a 50/50 or a variable beam splitter between the two object-beam mirrors, keeping in mind that the second

Reflection Hologram Setup (Two Object Beams)

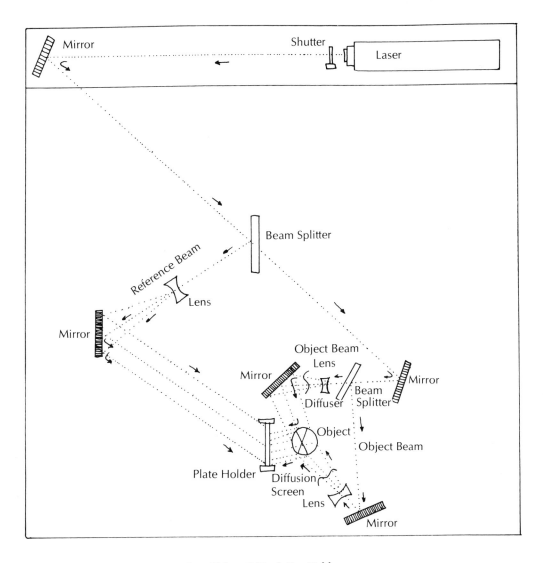

Overhead View Of Isolation Table

. . . . Laser Light Path
———————→ Direction of Laser Light Movement

object-beam path that comes into existence must be equal in length to the first.

6b. Place another mirror on the other side of the object so that the object is illuminated from both sides.

6c. Place one or two lenses in the path of this second object beam (or the 60× microscope objective). A diffuser can also be added for reducing glare from the illuminated object.

7. Place a 50/50 beam splitter (or a variable beam splitter) in the path of the laser light approximately in the middle of the isolation table. This other beam then serves as the reference beam.

8. Place a reference-beam mirror in the path of the reference beam and direct this light onto the cardboard screen in the plate holder. The reference-beam mirror can be placed so that the lengths of the laser light path are equal.

9. Measure the reference- and object-beam path lengths and adjust these distances so they are equal to within ½ inch. The object-beam path distance is measured from the beam splitter in the middle of the table to the object-beam mirror to the second object-beam mirror to the object to the cardboard screen in the plate holder. This distance should equal the reference-beam path distance, which is measured from the beam splitter in the middle of the table to the reference-beam mirror to the cardboard screen in the plate holder. (When two object beams are used, both object-beam path distances measured from the second beam splitter to the plate must be equal in length to themselves as well as the reference-beam path distance.)

10. Place one or two spreading lenses or a 60× microscope objective in the path of the reference-beam path so that the entire cardboard screen in the plate holder is evenly illuminated.

11. Measure the ratio of reference-beam intensity to object-beam intensity by using a light meter. In making a reflection hologram, the ratio of reference to object beam intensities should be on the order of 2:1 to 1:1, 1½:1 being about optimum. If the ratio is not the desired value, adjust the intensity by moving or removing the appropriate lens or lenses. When the ratio has been satisfactorily adjusted, consult the exposure chart to determine the appropriate exposure time. (The procedure for using a light meter is given in Section 3-2.1 "Recording a Transmission Hologram.")

12. Replacing the cardboard screen in the plate holder, take note of the resulting pattern. Lightly clean all surfaces of the components that encounter laser light by using lens tissue or isopropyl alcohol. Making sure all the components are firmly placed, leave the room for five to ten minutes (presettling time) to ensure the stability of the holographic components and the location in general.

13. Check the pattern on the cardboard screen with the earlier pattern. If the pattern appears to be stationary, turn off the overhead light and prepare to expose the plate.

14. Check the laser shutter to make sure it is working properly. Skip this step if the shutter is manually operated.

15. Remove the cardboard screen from the plate holder and any cardboard that remains from measuring the light ratio

(cardboard that blocks either reference or object beam). Haste and the anticipation of exposing the holographic plate at this point cause many would-be holographers to forget to check, resulting in no hologram.

16. Load the holographic plate or film (Agfa-Gevaert 8E75 NAH) in the plate holder with the emulsion side of the film facing the object. If there is antihalation backing on the plate, remove it before use (by soaking the plate in methanol) since reflection holograms cannot be recorded on plates or film that have backing on them.

17. Leave the isolation table area for ten minutes to allow the area to settle.

18. Expose the holographic plate to the laser light for the time determined by the exposure chart by turning on the shutter or manually exposing the plate.

19. Replace the exposed plate in a light-proof box and transfer it to the holographic processing area.

3-4.2 PROCESSING A REFLECTION HOLOGRAM

Processing a reflection hologram recorded on a silver-halide base holographic plate or film (for instance, Agfa-Gevaert 8E75 NAH plates or Kodak SO 173 film) consists of developing, stopping, and fixing the hologram in exactly the same way that a transmission hologram is processed. (For the necessary chemicals and darkroom supplies as well as step-by-step processing instructions, refer to 3-2.2—"Processing a Transmission Hologram.") The only difference between developing a transmission hologram and developing a reflection hologram consists in the degree to which a reflection hologram is allowed to develop. Whereas a transmission hologram is developed until there is approximately 60 per cent transmission (that is, when looking through the plate at a safelight, approximately 60 per cent of the light is transmitted, characterized by a light to medium gray), a reflection hologram is developed much darker, approximately 40 per cent transmission. In visual terms, the plate is about twice as dark as a developed transmission hologram; it is in fact, almost black in appearance.

It should be kept in mind that the darker the plate, the more silver is deposited in the recorded interference pattern and the more light can be reflected by the hologram during the reconstruction step of the hologram-making process. Care should be taken not to overdevelop the plate, though, since in overdeveloping, all contrast between the recorded interference pattern and the rest of the plate is lost, resulting in a low-efficiency hologram (in other words, a dim hologram). To further enhance the efficiency of a reflection hologram after processing, it is advisable to bleach the holographic plate. (For bleaching chemicals and procedure, refer to Appendix II (B). "Recipes for Bleaches.")

3-4.3 VIEWING A REFLECTION HOLOGRAM

The processed reflection hologram can be viewed either prior to or after bleaching. However, unlike the procedure for initially viewing a transmission hologram, the holographic plate or film should be completely dry in the case of a reflection hologram. This requisite exists because droplets of water on the surface of a wet holographic plate (instead of the recorded interference fringes of the hologram) would reflect any reference beam light re-illuminating the plate with a result of no visible holographic image. The procedure for viewing a reflection hologram follows.

1. Illuminate the reflection hologram with a white-light point source (the sun, a penlight, flashlight, slide- or movie-projector light), which acts as the reference beam in the reconstruction step of the process. When the hologram is handheld (instead of being mounted in a plate holder), tilt the plate until the holographic image appears, at which time light is striking the plate at exactly the same angle as did the reference beam in the recording. If the hologram is not bleached or treated with a triethanol-amine solution, the holographic image reflected into the viewer's eye will be a green color. The green is due to emulsion shrinkage and the ability of the hologram to filter green light selectively from the white light of the source.

2. To see the virtual image (usually the holographic image that appears in back of the plate), the emulsion side of the plate should be the side not illuminated by the white light. To see the projected image, the emulsion side of the plate should be illuminated.

3-5 SINGLE-BEAM REFLECTION HOLOGRAM SETUP

The single-beam reflection hologram set-up is probably the simplest and most economical (in terms of time expenditure in setting up and a minimum space requirement) of all known holographic camera configurations. This setup uses no beam splitter to divide the laser beam and requires no measurement of either beam ratio or object- and reference-beam paths since there is only one beam. However, the reflection hologram produced by this setup has a very small depth of field and a much smaller tolerance for movement of either object or holographic components. In essence, a single laser beam acts as both the reference and object beam. A spreading lens or microscope objective is used to spread the laser light so that the holographic plate is evenly illuminated (the reference beam). Laser light, which passes through the holographic plate, then illuminates an object (placed on the other side of the plate holder), which reflects light onto the plate (the object beam). A

Single-Beam Reflection Hologram Setup

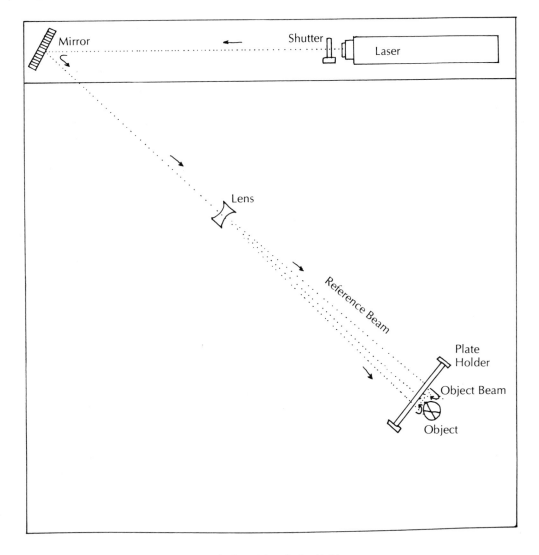

Overhead View Of Isolation Table

. . . . Laser Light Path
⟶ Direction of Laser Light Movement

listing of the holographic components required as well as step-by-step assembly instructions follows.

Holographic components used in a single-beam reflection hologram setup (not including the laser):

    1 laser transfer mirror
    2 spreading lenses
               or
    1 microscope objective (60×)
    1 plate holder
    1 cardboard screen
    1 light meter
    1 box holographic plates (Agfa-Gevaert 8E75 NAH)

Procedure for assembling a single-beam reflection hologram setup:

1. Mount the laser-transfer mirror in the corner of the table to direct the laser beam onto the isolation table.
2. Place the plate holder in the corner of the table opposite the laser-transfer mirror so that the laser beam illuminates the plate from an angle of about 30° (the angle most suited for convenient reconstruction; using other reference-beam angles makes it awkward for the viewer, since at such times the reconstructing reference beam and the viewer's eye lie in an approximate straight line).
3. Place the cardboard screen in the plate holder and center the point of laser light on the screen.
4. Place one or two lenses or the microscope objective in the path of the laser beam in order to spread out the light so that it illuminates the entire screen. Use a minimum of lenses since the intensity of light is effectively reduced by each lens (requiring long exposure times).
5. Place the object on a small pedestal or on a mound of sand on the side of the plate holder opposite the lens. The object should be positioned as close to the plate holder as possible without disturbing the plate or object when inserting or removing the plate before and after the exposure. When using a small power He-Ne laser (1 to 5 milliwatts) and a 4″ × 5″ plate or film, the optimum-size object is no larger than 3″ × 3″ × 4″ and either white or metallic in color so as to reflect the maximum amount of light. Make sure the object is completely stable since the slightest movement results in no hologram. Remove the cardboard screen before illuminating the object.
6. Measure the light-beam ratio by placing the photometer probe facing the laser (the reference beam) and facing the object (object beam). A ratio of 1:1 to 5:1 is acceptable; however, a ratio of 1:1 to 2:1 is optimum.
7. Leave the isolation table area for the standard presettling time period of about five minutes.
8. Check the laser shutter by turning it on and off. If it functions satisfactorily, turn the laser shutter off (or if manual, block the laser light with a cardboard hand-operated shutter).
9. Load the holographic plate or film (with the emulsion side toward the object) in the plate or film holder.
10. Leave the isolation table area for five to 10 minutes settling time.
11. Expose the holographic plate to the

laser light by turning the shutter on for the length of time prescribed by the exposure chart.

12. Replace the exposed plate in a light-proof box and process it according to the instructions for processing a reflection hologram (Refer to Section 3-4.2 "Processing a Reflection Hologram.")

3-6 IMAGE-PLANE HOLOGRAM SETUPS

Image-plane holograms are viewable with white light and are characterized by an orthoscopic image that is partially a virtual image and partially a real or projected image. In other words, the image-plane hologram can be made so that its holographic image projects either in back of the holographic plate, in front of the holographic plate, or partly in back and partly in front. The placement of the holographic image is completely up to the discretion of the holographer.

There are at least several different types of holographic camera setups for making image plane holograms. The simplest setup consists of a standard transmission or reflection setup in which an additional lens is placed between the three-dimensional subject and the holographic plate holder. This additional lens is positioned so that the subject is at one focal plane of the lens and the holographic plate is at the other focal plane. The lens itself gathers all the light from the subject and focuses it in the plane of the holographic plate. This object-beam light together with the reference beam forms an image plane hologram made by using a lens.

The holographic image of such a one-step image-plane hologram is limited by the size of the lens used to gather the light from the subject. The lens itself acts as an aperture in effectively inhibiting both vertical and horizontal parallax. What happens is that the resulting holographic image can't be seen from as many sides and angles as the image from a standard transmission or reflection hologram. The lens actually reduces much of the three-dimensional look-around qualities of the image.

The holographic image of a one-step image-plate hologram, however, can be seen by using white light such as sunlight or the light from a straight-filament light bulb. Laser light is unnecessary to re-illuminate the hologram. The holographic image of such a hologram is called an achromatic image, which literally means that the image itself is without color. Since the holographic image of an image-plane hologram is focused exactly onto the plane of the holographic plate, there is no color blurring or spectral smearing, which occur when a standard transmission hologram is illuminated by white light. Instead, all the points of light that constitute the image retain their sharpness, resolution, and recognizability so that a three-dimensional image which is exactly focused becomes apparent.

One-Step Transmission Image-Plane Hologram Setup (Using A Lens)

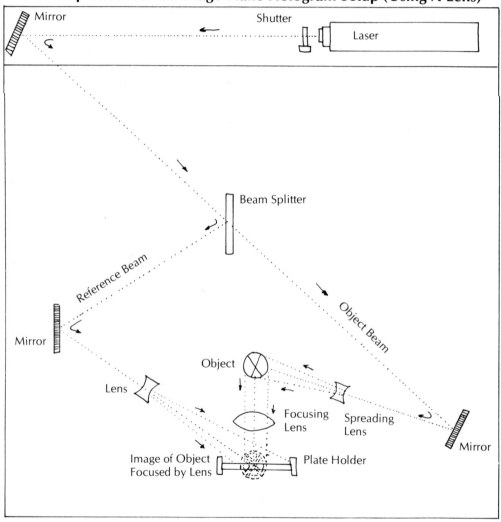

Overhead View Of Isolation Table
. . . . Laser Light Path
⟶ Direction of Laser Light Movement

One-Step Reflection Image-Plane Hologram Setup (Using A Lens)

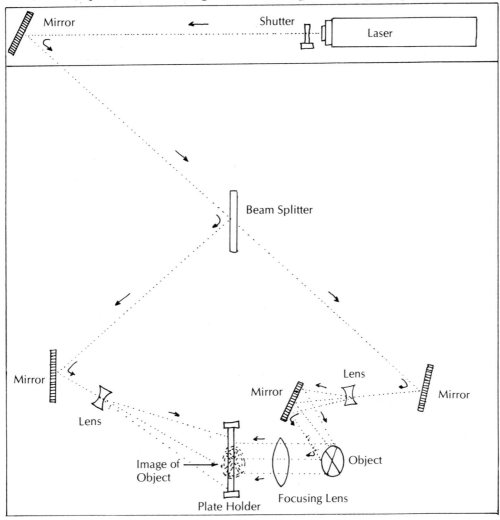

Overhead View

.... Laser Light Path

——➤ Direction of Laser Light Movement

One-Step Transmission Image-Plane Hologram Setup (Using A Spherical Mirror)

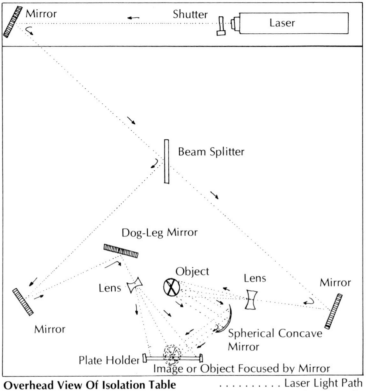

Overhead View Of Isolation Table   . . . . Laser Light Path
                                              ⟶ Direction of Laser Light Movement

Another one-step method of making an image-plane hologram consists of using a spherical mirror instead of a lens to gather the light from an illuminated subject and then focus it onto the plane of the holographic plate. This focused light, which is the object beam, interferes with a reference beam, producing an image-plane hologram.

The concave and spherical mirror used in the setup must be large enough to gather all the light from a laser-illuminated subject. This type of mirror is difficult to find, though surplus stores usually carry them, and rather expensive. Also, this method is unsatisfactory for making holograms of more than one object at a time since the mirror can only project so much information. The image of such a hologram must be exactly centered on the holographic plate or film and must be restricted to a single small object when $4'' \times 5''$ plates or film and a laser of a couple of milliwatts are used.

Another technique for making image-plane holograms is the two-step method. The first step consists of making a master-plate hologram—simply a standard transmission hologram with two object beams for more uniform illumination. In

One-Step Reflection Image-Plane Hologram Setup (Using A Spherical Mirror)

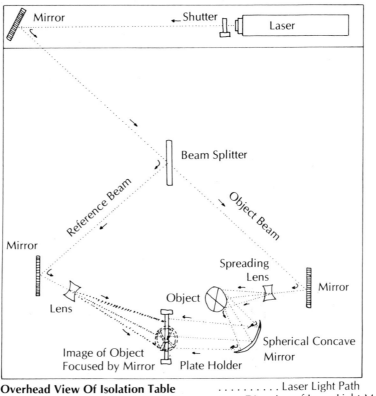

Overhead View Of Isolation Table   ·········· Laser Light Path

⟶ Direction of Laser Light Movement

the second step of the process, the projected image from the master hologram is used as the subject for the image-plane hologram. The projected image of the master hologram is positioned so that it intersects the plane of the holographic plate or film. With the addition of a reference beam, either a transmission image-plane hologram or a reflection image-plane hologram is formed. In a sense, this type of image-plane hologram is a hologram made from a hologram since a holographic projected image, instead of a real three-dimensional object, is used as the subject.

The two-step type of image-plane hologram is a much more satisfactory method than the one-step method because it allows a greater degree of precision in choosing the exact spatial location of the image. Also, the holographic image itself is not reduced as it is when a lens or a mirror is used to focus the light onto the holographic plate. In a sense, by using the two-step method of making an image-plane hologram, the hologram itself, which produces the projected image, acts in the same capacity as the lens or mirror in the one-step method.

Recording A Two-Step Image-Plane Hologram (Rainbow Hologram)

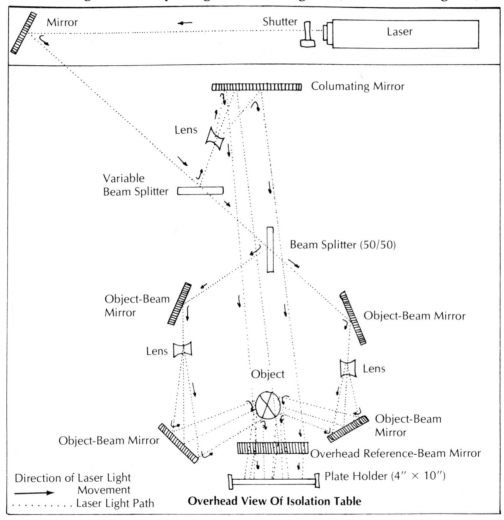

Step 1: Recording A Master-Plate Hologram (A Transmission Hologram With Two Object Beams And An Overhead Reference Beam)

3-6.1 RECORDING A MASTER-PLATE HOLOGRAM

Making a master-plate hologram is essentially the same as making a standard transmission hologram with two object beams. There are, however, a few modifications in the actual setup and the holographic technique, which produce more satisfactory results. First, the reference beam should be collimated so that plane instead of spherical light waves illuminate the holographic plate. The reference beam can be collimated by using a collimating mirror, such as a telescope mirror, for the reference-beam mirror. Collimated laser light is preferred because there will be no distortion of the projected image during the reconstruction step of the process. Since the projected image is actually the subject of the image-plane hologram, distortion of the image would be highly undesirable.

Because the size of the holographic image is directly dependent upon the size of the master plate, the size of the master plate should be preferably larger than the size of the image-plane hologram that is produced from it. When a 4″ × 5″ holographic plate is used for the master plate as well as the image-plane copy hologram, the size of the holographic image is limited to about 2″ × 3″. In other words, to have an undistorted projected image from the master plate, the subject for the master plate should not exceed these dimensions. When the subject is larger, the holographic image of the image-plane hologram may become partially cut off since the holographic plate is not large enough to accommodate the entire image. Since only a portion and not the whole of the image is visible, the entire holographic image loses its three-dimensional quality.

To eliminate the undesirable distortion effects, the master-plate hologram should be considerably larger than the image-plane holograms, 4″ × 10″ master Holograms™, two-step transmission image-plane holo-grams, 4″ × 10″ master plates were used. These master plates were made by cutting the standard available 8″ × 10″ holographic plates in half. White-light-viewable copies were then made on 5-inch or 10-inch Agfa-Gevaert 10E75 film cut into pieces measuring about 5″ × 10″. However, 4″ × 5″ holographic plates have been used for both master plates as well as copies. The size of the resulting holographic image was greatly restricted in this case.

Another modification of the standard transmission hologram setup utilized by the School of Holography in the Rainbow Hologram Camera is the use of an overhead reference beam instead of a reference beam that illuminates the holographic plate from the side. Because of this change, reconstruction of the holographic image can take place when the Rainbow Hologram™ is re-illuminated by an overhead lamp, such as a reading lamp. This modification allows uncomplicated viewing of the Rainbow Hologram™ from commonly available sources of white light.

3-6.2 RECORDING AN IMAGE-PLANE HOLOGRAM

The second step in the procedure consists of using the projected image from the master plate as the object beam in the image-plane hologram. This is done by positioning the master plate so that when it is

Recording A Two-Step Image-Plane Hologram (Rainbow Hologram)

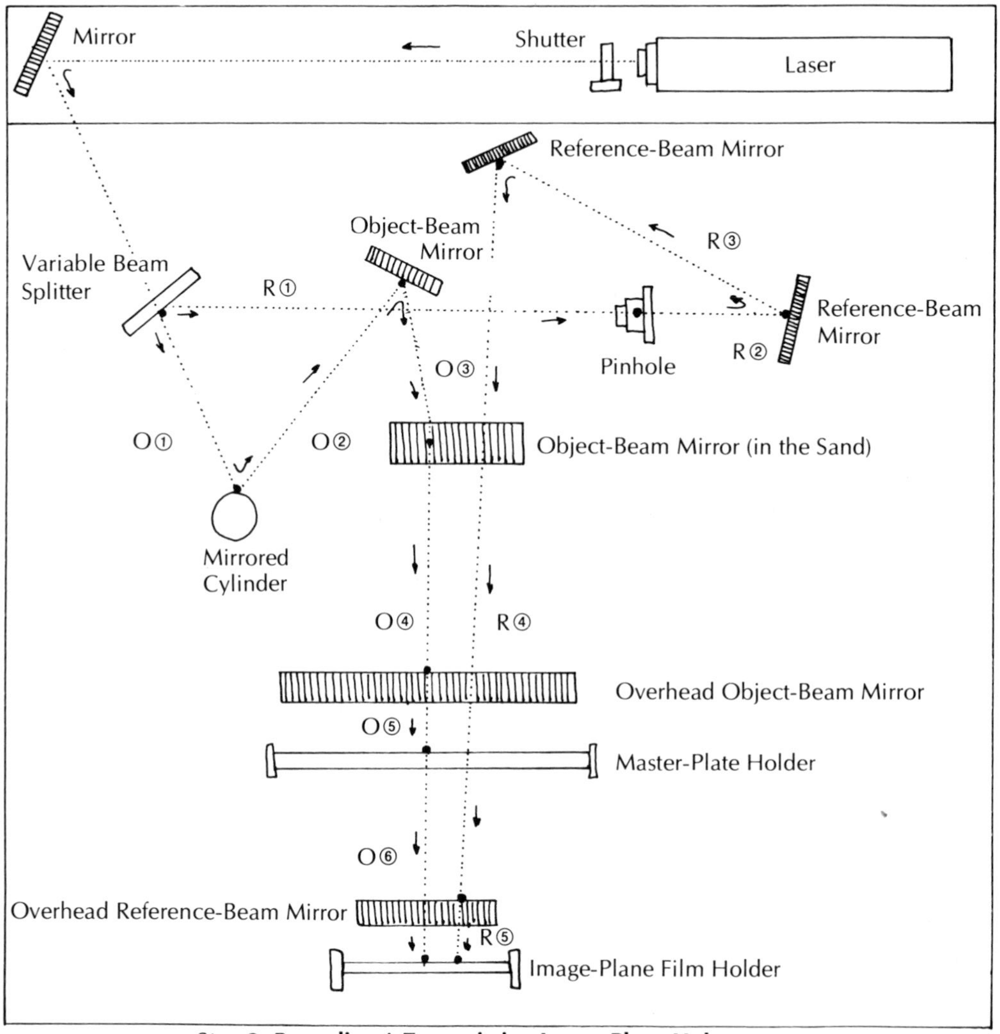

Mirror · Shutter · Laser

Reference-Beam Mirror

Object-Beam Mirror

R③

Variable Beam Splitter · R①

Reference-Beam Mirror

R②

Pinhole

O③

O① · O②

Object-Beam Mirror (in the Sand)

Mirrored Cylinder

O④ · R④

Overhead Object-Beam Mirror

O⑤

Master-Plate Holder

O⑥

Overhead Reference-Beam Mirror

R⑤

Image-Plane Film Holder

Step 2: Recording A Transmission Image-Plane Hologram
. . . . Laser Light Path
→ → → → Direction of Laser Light Movement
°① • . . . • Optical Path ("O" signifies object beam; "R" signifies reference beam.)

re-illuminated by laser light, the projected image is produced. The space that contains the projected image is then intersected by the plane of the adjustable film holder. The position of the adjustable film holder determines whether the resulting image of the image-plane hologram will be in front of the film, in back of the film, or partially projecting in back and in front of the film. Depending on which side of the film is illuminated by the reference beam, either a transmission image-plane hologram or a reflection image-plane hologram is formed.

The Rainbow Hologram Copy Camera, which produces the image-plane holograms, also uses a cylindrical lens in the object beam. When laser light passes through a cylindrical lens or is reflected from a mirrored cylinder, the point of laser light is spread out to a narrow horizontal band of light. This horizontal band of laser light illuminates the master plate and causes the projected image to appear. When all the object-beam light is channeled in this way, instead of illuminating the entire master plate, the resulting image-plane copy hologram is relatively much brighter. This is due to the fact that all the intensity of the laser light remains in this single horizontal band of light instead of being dissipated by an even but overall dimmer illumination. Also, through the use of the cylindrical lens, vertical parallax is mostly eliminated; however, complete horizontal parallax remains intact.

The optical paths of laser light in the holographic copy camera setup are as follows: The single beam of light that emits from the laser is reflected from the laser-transfer mirror in the corner of the table onto the isolation table surface. Then, the laser beam encounters a variable beam splitter that divides it into the object beam and the reference beam. The object beam passes onto a mirrored cylinder that acts as a cylindrical lens, thereby making the beam of laser light a horizontal band of light. This horizontal band of laser light is reflected onto a small mirror and then onto a mirror in the sand. This mirror in the sand reflects the light to an overhead mirror. The overhead mirror reflects the horizontal band of light onto the master plate from the same angle as the reference beam was in the master plate. This same angle is the *angle of reconstruction*, which produces the projected image.

Simultaneously, the reference beam is reflected from two mirrors, passes through a pinhole to another large mirror, and then is finally reflected onto another overhead mirror. This overhead reference-beam mirror then reflects the laser light onto the film in the adjustable film holder.

A listing of the holographic components necessary for the assembly of a transmission image-plane hologram setup, the *image-plane copy camera*, as well as the actual procedure follows.

Holographic components used in a transmission image-plane hologram setup (not including the laser):

> 5 mirrors ($2'' \times 3''$)
> 4 mirrors (at least $6'' \times 10''$)
> 1 variable beam splitter
> > *or*
> 1 beam splitter (50/50)
> 1 cylindrical lens
> > *or*
> 1 mirrored cylinder

1 diffusion screen (for the image-plane film holder)

1 master-plate holder (to hold either a 4″ × 5″ or 4″ × 10″ master plate)

1 image-plane film or plate holder (for a 4″ × 5″ or 5″ × 8″ plate or film)

1 cardboard screen (to fit the image-plane film holder)

1 pinhole or other spatial filter

1 tape measure

1 box of holographic plates or film (Agfa-Gevaert 8E75)

Procedure for assembling a transmission image-plane hologram setup:

1. Place a laser-transfer mirror in the corner of the table near the laser in order to direct the laser beam onto the table.

2. Place the master-plate holder on the table on the side opposite the laser and approximately parallel to the position of the laser. Leave about 6 inches of space from the master-plate holder to the edge of the table. This space will later be used for mounting the image-plane film holder.

3. Place the variable beam splitter in the path of the laser light approximately in the same position as the beam splitter in a transmission hologram, maybe even several inches closer to the laser-transfer mirror.

4. The next several steps in the procedure describe mounting the components of the object-beam path, from the beam splitter to the master plate holder. Place the mirrored cylinder several inches from the beam splitter.

5. Mount a small mirror approximately parallel to the laser in the path of the object beam. The line of light produced by the cylinder will reflect from it.

6. Mount another mirror in the sand so that the horizontal line of laser light will reflect upward at an angle to the larger overhead mirror.

7. Mount the overhead mirror directly above the master-plate holder so that the horizontal band of laser light illuminates the master plate from the same angle as the reference beam did in the master-plate camera. This horizontal band of light will reconstruct the wave fronts that make up the projected image of the master plate. Simultaneously, it acts as the object beam for the image-plane hologram. The overhead mirror can be mounted on plastic or aluminum tubing like the other components.

8. Adjust the overhead mirror so that the reconstructing horizontal band of laser light illuminates the master plate in the master-plate holder, thereby producing the projected image. This step completes the installation of the object-beam components.

9. The next several steps detail the installation of components that direct the reference beam from the variable beam splitter to the adjustable film holder. First, place the adjustable film holder on the table so that the projected image from the master plate intersects the plane of the film holder. The diffusing

Optical Path Distances Of The Transmission Image-Plane Hologram Setup

Reference Beam

R①	Beam Splitter to Front of Pinhole	16″
R②	Pinhole to 1st Reference-Beam Mirror	5½″
R③	1st Reference-Beam Mirror to 2nd	11½″
R④	2nd Reference-Beam Mirror to 3rd	34″
R⑤	3rd Reference-Beam Mirror to Image-Plane Film Holder	9½″
	Total	76½″

Object Beam

O①	Beam Splitter to Cylinder	10″
O②	Cylinder to 3″ × 4″ Mirror	16
O③	3″ × 4″ Mirror to Mirror in the Sand	10″
O④	Mirror in the Sand to Curved Plastic Overhead Mirror	21″
O⑤	Overhead Mirror to Master Plate	11½″
O⑥	Master Plate to Image-Plane Film Holder	9″
	Total	77½″

screen placed in the adjustable film holder facilitates locating the projected image. The film holder is made adjustable by mounting it on a movable track or simply by clamping the plate holder to a wooden baseboard with a C-clamp. In this way, the plate holder can be moved and adjusted for whatever position is desirable.

10. Mount the overhead mirror for the reference beam. This mirror can either be free-standing or it can be mounted on a wooden beam that spans the sand table. The mirror should, in any case, be adjustable to allow for the different reference-beam angles resulting from the changing position of the adjustable film holder.

11. Mount another large mirror approximately parallel to the laser and centered in relation to the table. This mirror reflects the light of the reference beam from a smaller mirror to the overhead reference-beam mirror.

12. Mount another smaller mirror to transfer the reference beam from the beam splitter to the other large mirror, thereby completing the dogleg optical path of the reference beam.

13. Measure the distances of the reference- and object-beam optical paths and adjust the components so that the distances are equal to within an inch. Because the image-plane film holder can be moved, the distances of both paths will not be exactly equal, even though a tolerance of not more than several inches is acceptable.

14. When the distances of the two optical paths are measured to satisfaction, mount the pinhole or spatial filter in the path of the reference beam. The pinhole should be placed between the beam splitter and the smaller mirror. The pinhole filters the laser light so that dust patterns and other undesirable effects are not recorded in the hologram.

15. Adjust the ratio of light intensities by eye. In other words, adjust the beam splitter until the reference beam intensity is about equal to that of the object beam. Check the intensity with a light meter in order to determine the exposure time.

16. Leave the isolation table area for about ten minutes (presettling time).

17. Check the laser shutter to make sure it is working properly.

18. Turn off the laser shutter, or block the laser beam with a piece of cardboard in the case of manual shutter operation.

19. Load the holographic plate or film in the adjustable film holder.

20. Leave the isolation table area for ten minutes settling time.

21. Expose the holographic film to the laser light for the time duration determined by the exposure chart.

22. Place the exposed film in a lightproof box and process it according to the instructions for processing a hologram.

3-6.3 VIEWING A RAINBOW HOLOGRAM™

A Rainbow Hologram™ can be seen with almost any undiffused point source of white light. When sunlight, full moonlight, or candlelight illuminates the hologram, a holographic image appears in each of the colors of the rainbow, as the name implies. Each of the different colors can be seen by changing the position of the eye vertically in relation to the hologram. By viewing the hologram and scanning it horizontally from side to side, the three-dimensional qualities of the rainbow holographic image can be seen.

A straight, bare-filament light bulb, however, is probably the ideal light source to use for wave-front reconstruction of the holographic image. Either a 60- or 100-watt vertical or horizontal filament light bulb can be used. The light bulb should be clear or unfrosted, though, since a frosted light bulb emits diffused light that is unusable for this purpose.

The Rainbow Hologram™ should be mounted at eye level so the light bulb that illuminates the hologram is above and in back of the hologram at about a 45° angle. In this position, optimum conditions for viewing a Rainbow Hologram™ are fulfilled.

APPENDIXES

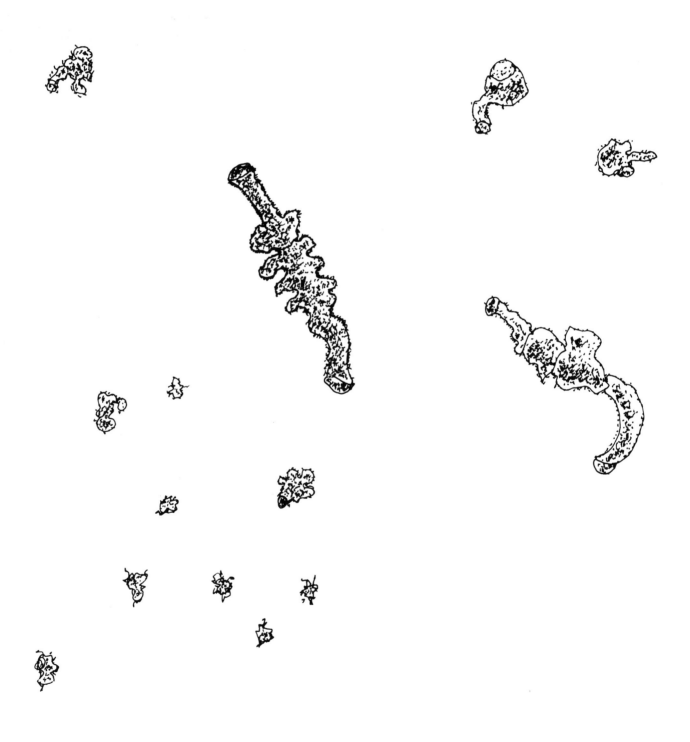

APPENDIX I (A).   DESCRIPTION AND THEORY OF HELIUM-NEON (He-Ne) LASERS

In order to understand how a laser operates, it is necessary to be aware of the energy exchanges that naturally occur within an atomic system. These exchanges of energy involve either absorption or emission of electromagnetic energy within a single atomic system or a transfer of energy from one atomic system to another.

Ordinarily, atomic systems exist in an energy state or at an energy level where the energy is lowest. The atomic system remains stable and utilizes the least amount of energy to continue doing so. This energy state or energy level is called the *ground state*, or the *ground level*. The ground state corresponds to an atomic system in which the electrons orbiting the nucleus occupy space in such a way that the least amount of energy is expended to maintain the same condition. The only possible transfer of energy that an atomic system in the ground state can undergo is *absorption of energy*.

The absorption of energy within an atomic system takes place when energy in the form of a photon interacts with an electron and causes the energy level to be raised. Energy absorption also takes place when an electron is excited by direct electrical discharge. An electric field accelerates the electrons, and the resulting collisions between electrons raise the energy level. This higher energy level is called the *excited level*, or *excited state*.

An atomic system, however, can be excited only by specific and precise amounts of energy called *quantums* of energy. The exact amount of quantum energy varies for each atomic system as well as for each of the energy levels inherent in the system. In other words, only photons or electrons of a specific wavelength or energy can raise an atomic system from its ground level to an excited level.

When an atomic system is in an excited state, this means that an electron orbiting around the nucleus will be found in a different electron shell than when the system is at ground level. A different elec-

Transfer Of Energy In An Atomic System

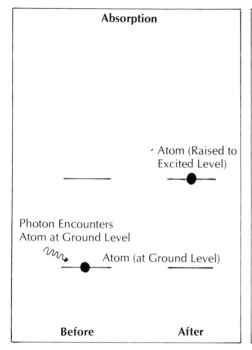

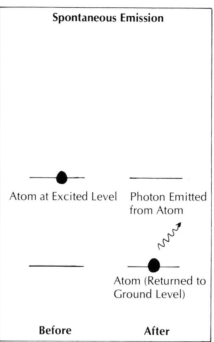

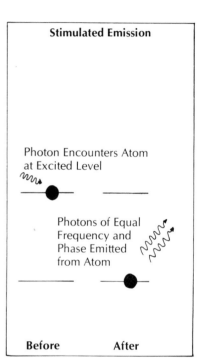

tron shell refers to a different space than the one usually occupied by the electron. The simultaneous though brief occupation of space by an excited electron and an unexcited electron is characteristic of an excited energy level.

There is a natural tendency for an atomic system to exist in the lowest possible energy state. Because of this tendency, an atom that is excited into a higher energy level may return directly to the ground level, or a lower energy level. The return of the atomic system to a lower energy level may take place with the release of energy in the form of electromagnetic radiation. The release of a photon in this case is called *spontaneous emission*. The energy contained in the released photon is exactly equal to the difference in energy levels between the excited state and the lower energy state. In other words, only certain frequencies of electromagnetic energy can be absorbed by certain atomic systems and, likewise, only certain frequencies can be spontaneously emitted.

Besides spontaneous emission of energy, transitions between different energy levels can take place due to stimulation or excitement of the atomic system. These so-called *stimulated emissions* of electromagnetic energy are released simultaneously with the spontaneous emissions. The difference between the two types of emission, however, is that the photon released during stimulated emission is identical in frequency, phase, and direction to the photon that acts as the stimulation. In other words, a photon released by an excited atom can trigger another excited atom into emitting an identical photon.

Spontaneous emission produces light that is more or less random in frequency and direction. Light from spontaneous emission is the natural output of ordinary light sources. White light, for example, appears white to the human eye because there are enough photons of different frequencies to produce this effect. Stimulated emission, however, produces light that is coherent, is a single frequency or single color, and is in phase—in other words, *laser light.*

Stimulated emission usually does not produce laser light simply because there are not enough atoms in the excited state in any given space and time. Since atomic systems tend to remain in the ground state, stimulated emission is negligible until there is a *population inversion.* A population inversion means that the population of atoms in the ground state is inverted or changed so that they are all raised to a higher energy level. When population inversion takes place, lasing action becomes possible.

Another requirement for the production of laser light is an enclosed space, called an *optical cavity,* containing the pre-excited atoms, which are collectively called the *lasing medium.* The optical cavity is usually a glass, ceramic, or aluminum tube with sealed-off ends. The optical cavity not only physically contains the lasing medium in a confined space, but also provides direction for the photons that can travel throughout the length of the laser tube. The laser tube is bounded on both ends by laser mirrors, one of which is usually more reflecting than the other. When population inversion is established and retained due to continuous excitation, lasing action takes place and a laser beam of light emits from one end of the laser tube. *This lasing action is the result of photons passing from one end of the tube to the other, reflecting from the end mirror, stimulating the creation of more photons by exciting more atoms, which emit more photons, and so on.*

The word "laser" is an acronym for Light Amplification by Stimulated Emission of Radiation. In electronic terms, a laser is essentially an oscillator at visible electromagnetic frequencies. This oscillator simply amplifies light energy by means of a feedback circuit. The laser tube, which is also called an optical resonator or resonant cavity, acts as the amplifier while the two laser mirrors act as the feedback circuit. Losses in the feedback circuit occur when photons pass through the end mirrors. When photon losses are overcome by the effects of amplification within the resonant cavity, electromagnetic oscillations are set up throughout the length of the laser tube. When the oscillations of energy are established in a state of equilibrium, or balance, they are emitted from the laser tube as an intense beam of single-frequency light, the one and only laser beam.

Because of the relatively low cost and relative ease in construction and maintenance, helium-neon lasers are un-

Helium-Neon Laser Energy-Level Diagram

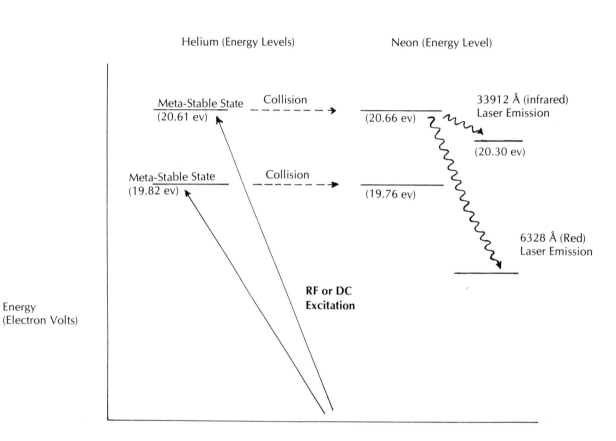

Ground States Or Ground Levels

Helium-Neon Laser

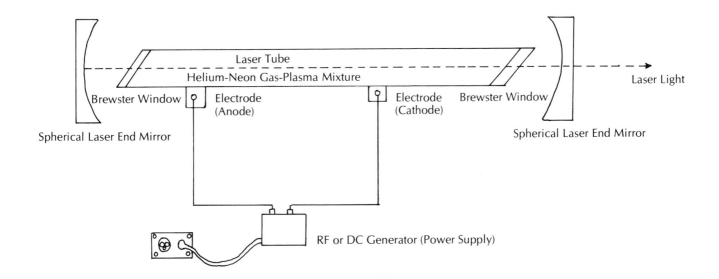

Side View Diagram

doubtedly the most commonly used lasers in making holograms. The lasing medium is a helium and neon gas plasma, a mixture of about 90 per cent helium and 10 per cent neon gases. These gases are contained in a sealed tube sometimes called a plasma tube. The laser mirrors at both ends of the tube are either plane, spherical, or a combination of plane mirrors and a prismatic wavelength selector. A prismatic wavelength selector is simply a prism that can be adjusted to reflect either the 6328 angstroms red or the 3.39 infrared wavelength. The laser tube itself has *Brewster windows* mounted on both ends. A Brewster window usually consists of a flat piece of quartz mounted at the Brewster angle, which is about 55° for quartz at 6328 Å. The Brewster angle is used because laser light or any other light can be entirely transmitted as a polarized beam of light after it passes through a surface tilted at this angle. This is especially useful since the Brewster windows effectively eliminate any losses in intensity due to internal reflections inside the laser tube.

Excitation of the gas plasma is initiated by an electrical discharge in the form of DC current or RF stimulation. In a He-Ne laser, the energy of the electrical discharge raises the energy of the helium atoms to the so-called *metastable state*. The metastable state is an energy level that is higher than the ground level and where excited atoms tend to remain longer than in another excited state. In the metastable state, the excited atoms of helium give up their energy to the unexcited neon atoms during electron collisions. The energy of the electrical discharge, after residing in the energized state of helium, passes onto the neon atoms when energy is transferred. When population inversion occurs among the neon atoms, there will be more excited neon atoms than unexcited ones. The excited neon atoms dissipate their energy by making a transition to a lower energy state. During this transition of energy, electromagnetic radiation will be emitted as the visible red wavelength of 6328 angstroms.

Once stimulated emission of photons is initiated, the reflecting mirrors at both ends of the laser tube are used to obtain *continuous wave oscillations* of laser light. A continuous wave laser refers to a laser that emits light radiation more or less continuously in relation to time and space. The *pulse* mode of operation refers to a laser that emits pulses of radiation lasting only nanoseconds—minute fractions of a second in time. The He-Ne laser, which emits 6328 Å red, is a continuous wave laser.

It should be noted that there are many different energy levels to which the excited neon atoms can turn and as many different and separate wavelengths of energy that can be produced. Most of the wavelengths that have been observed in a He-Ne laser are in the infrared band of radiation, and most of these have a very low power output. However, because of the position of the laser mirrors, the He-Ne laser has been optimized in the production of visible red laser light. Its popularity in making holograms is due only to the relatively low cost of producing coherent red light.

APPENDIX I (B). OTHER LASERS

Besides the helium-neon laser, there are other lasers that are used to make holograms. These can be described according to their lasing media: gas lasers, solid-state lasers, and liquid lasers.

Gas lasers generally are lasers that utilize the noble (inert) gases: helium, neon, argon, krypton, xenon, and their ions. The gases are used either singly or as a mixture. The most common lasers of this group are the ion lasers, which can be operated either in the continuous wave or the pulsed mode. The lasing medium in the so-called ion laser is argon gas, krypton gas, or a mixture of the two gases, depending on what wavelengths of light are desired. A krypton-gas ion laser produces lasing action at wavelengths ranging from red (6471 Å), yellow (5682 Å), green (5208 Å), to a blue-violet (4762 Å).

Excitation of the gas into its ionic state is accomplished by a DC electrical discharge. First, the gas enters into its unexcited ionic state and then into the excited higher energy level from which the ion has the capacity to provide lasing action.

Unlike helium-neon lasers, a krypton ion laser is bulky, requires an elaborate power supply and power requirements, and must be used only with a water-cooling system that neutralizes the heat generated during lasing. Also, because of the many available wavelengths, wavelength selection is made by either changing the laser mirrors, using the prismatic wavelength selector, or manually using the fine-tuning controls.

The argon ion laser produces laser light mostly in the green and blue end of the visible spectrum: 5145 Å green, 4880 Å blue, and a violet line, 4579 Å. Compared with its krypton laser counterpart, the argon laser is less bulky and doesn't require water cooling when the power output is one watt or less. The cost of an argon ion laser, however, is still several times greater than the cost of a helium-neon laser.

Apart from the gas ion lasers, the so-called *metal vapor lasers* are also finding increased usage in holography. The metal vapor lasers are the helium-cadmium (He-Cd) laser and the helium-selenium (He-Se) laser that produce laser light in the deep blue (4416 Å) and the ultraviolet (3250 Å) end of the electromagnetic spectrum.

The plasma tube of the helium-cadmium laser is filled with a mixture of helium gas and cadmium vapor, which is produced by heating about 3000 mg of solid cadmium to its vaporized state. The excitation of the plasma with DC current is responsible for certain energy transitions that result in the emission of laser light.

Besides making holograms with the visible blue laser light, the ultraviolet wavelength of a helium-cadmium laser can be used in making copies of holograms. In effect, copies of a hologram can be made by photographically reproducing the re-corded interference pattern with a type of contact-printing. The light used in this reproduction can be ultraviolet laser light. Since ultraviolet laser light is of a higher frequency than the red helium-neon laser light, it has more inherent energy, a greater power density, and more efficiency for reproducing holograms.

Solid-state lasers are lasers in which the lasing material is embedded in a solid. The ruby laser uses a rod of aluminum oxide (Al_2O_3), which is also called synthetic sapphire. The aluminum oxide rod is doped with about .05 per cent chromium oxide ($Cr_2 O_3$). *Doped,* in this sense, means that a very small amount of one substance is added to another relatively pure substance. The chromium impurity within the rod gives the ruby rod its red color and the chromium ions(Cr^{+++}) are the active ingredients that produces the lasing action.

A xenon flash lamp is usually coiled around or near the cylindrical ruby rod. The flash lamp provides the light and excitation energy necessary to raise the electrons of the chromium atoms to an excited energy level. With population inversion of the atoms, stimulated emission occurs at the wavelength 6934 Å red. When amplification takes place, the ruby laser emits light in an intense pulse that lasts only several microseconds.

Because of the strobelike quality, which effectively stops motion, pulsed ruby lasers have been used to make holograms of fast-moving subjects such as insects, aerosol particles, and humans. Holograms of human subjects made with pulsed ruby lasers, however, have a tendency to appear waxy and very much unlifelike due to the unnatural lighting of high-intensity red light.

Other solid-state lasers besides the

ruby laser include the Nd-YAG laser (neodymium– yttrium aluminum garnet) and the Nd– glass laser in which a glass rod is doped with a small quantity of neodymium. The YAG laser produces several wavelengths of pulsed infrared radiation (1.06 and 1.34 micrometers) at several watts, and somewhat less in the continuous wave mode.

In addition, the YAG laser produces red laser light (6700 Å) and also green laser light (5320 Å) when a frequency doubler is coupled to it. The *frequency doubler* consists of a crystal of Barium Sodium Niobate or some other material that generates a second harmonic of the 1.06 micrometer wavelength. The second harmonic of this infrared frequency is green laser light.

Liquid lasers, which are usually called *organic dye lasers,* can be tuned to any wavelength from about 4000 to about 7500 Å. Dye lasers use either an argon ion laser, a pulsed nitrogen (N_2) laser, or a flash lamp to excite a dilute solution of organic dye. These dyes include rhodamine 6G, which produces laser light from yellowish-green to red; 7-diethylamino-4-methylcoumarin, which can be tuned from 4300 to 4900 Å; sodium fluorescin, which produces laser light around 5500 Å green; and others.

When an organic dye in an alcohol solution is stimulated by laser light or by the intense white light of a flash lamp, the dye molecules absorb enough light energy to fluoresce, or give off light. After continued reflections from the laser mirrors, enough energy is developed so that the light begins to lase. A prism or a filter can be used to tune the desired wavelength of light selectively and continuously. The dye solutions usually are enclosed in cassettes so that the dyes can be changed quickly and effortlessly. Dye lasers, however, are not used much in holography.

APPENDIX II (A). THEORY OF DEVELOPMENT AND SUBSEQUENT PROCESSING (SILVER-HALIDE EMULSION)

The most common light-sensitive recording material used in photography and holography is the *silver-halide emulsion.* The silver-halide emulsion consists of submicroscopic crystals of a silver halide, such as silver bromide or silver iodide, suspended in gelatin and coated onto a substrate, such as plate glass or a film base.

The thickness of the entire emulsion is usually on the order of 3 to 12 micrometers, only 3 to 12 ten-thousandths of a centimeter. The gelatin itself acts as the medium that holds the silver-halide crystals in place and also controls the size of the crystals by its very presence. A gross model of the situation inside the emulsion can be represented by an evenly spread layer of solidified Jello on a piece of white bread. The Jello represents the gelatin suspension and the bread represents the substrate.

When a silver-halide light-sensitive emulsion is exposed to light, one quantum unit of energy, or one photon, is absorbed by each silver-halide crystal that is in the path of the light. Each crystal that undergoes contact with the light energy becomes marked. Collectively, these marked silver-halide crystals are called the *latent image*. After exposure to light energy, the latent image is developed and subsequently processed so that the *visible image* appears.

The exact mechanism by which the latent image is formed is not completely known. However, the steps taken to process the exposed film are standard procedure. Photographic processing means converting the latent image into the visible image, by treating the exposed film with various chemicals. Because the chemicals used to treat the exposed film are in liquid form, this chemical treatment is called *wet processing.*

Developing, which is the first step of wet processing, consists of the reduction, or change, of the exposed silver-halide crystals into grains of silver under the influence of a developing agent. When this reaction has thoroughly taken place throughout the emulsion, silver metal—as grains of silver—is deposited in the spaces that mark the latent image. The deposited silver then reflects light into the eye of the viewer, producing the visible image.

The procedural steps in wet processing are developing, stopping, fixing, and bleaching, with washing in between each step. *Stopping* consists of immersing the developed film into a stop bath. The stop bath, which is also called an acid rinse bath, is used to stop the precipitation of silver initiated by the developer. The stop bath is simply an acetic acid solution that instantly neutralizes the action of the developer.

The *fixing* bath is used next to fix the precipitated, or deposited, silver in place. Fix also washes out any undeveloped silver-halide crystals and hardens the gelatin emulsion. The usual ingredient of the fixing bath, ammonium thiosulfate, acts to draw all the undeveloped silver compounds out of the emulsion. If hardening is desired, then Rapid Fix (suggested earlier) must not be used as it does not contain hardener.

Bleaching the film or plate in the last step of wet processing again reconverts the precipitated silver into a transparent silver salt. This transparent silver salt is found exactly in the spaces occupied by the original latent image. This transparency is important in holography because a bleached film or plate tends to diffract light more efficiently than an unbleached hologram, and a hologram that diffracts more light is a brighter hologram.

APPENDIX II(B). RECIPES FOR BLEACHES USED ON SILVER-HALIDE HOLOGRAPHIC PLATES

The following bleaches can be used to bleach transmission holograms recorded on Agfa-Gevaert or Kodak holographic plates:

CAUTIONARY NOTE: In the formulas below, wherever directions call for a mixture of water with acid (especially concentrated sulfuric acid), *always* add the acid to the water, not the other way around. Pour the acid slowly and carefully into the water so that it does not splash. The sulfuric, in particular, is highly concentrated and will burn you.

Ferricyanide Bleach
1. Mix together:

Potassium ferricyanide $[K_3Fe(CN)_6]$	45 g
Potassium bromide $[KBr]$	20 g
or	
Potassium iodide $[KI]$	25 g
Water	1 liter

2. Bleach till clear

Ferric Bleach
1. Mix together:

Ferric bromide $[FeBr_3 \cdot 6H_2O]$	70 g
Ferric chloride $[FeCl_3 \cdot 6H_2O]$	50 g
and	
Sulfuric acid $[H_2SO_4]$	4 ml
Water	1 liter

2. Bleach till clear, then place in water for one minute.
3. Place bleached plate in isopropyl alcohol for one minute.

Cupric Chloride Bleach

1. Mix together:

Water at 52 C	750 ml
Cupric chloride [CuCl$_2$]	10 g
Citric acid	10 g

2. Add water to make   1 liter
3. Bleach till clear

Two-Step Bleach

1. Bleach in 5% solution of cupric bromide [50 g CuBr$_2$ in 1 liter of H$_2$O]
2. Rinse in water
3. Clear in solution of 1 part Solution A to 10 parts Solution B.

Solution A:

Potassium permanganate [KMnO$_4$]	5 g
Water	1 liter

Solution B:

Sulfuric acid	10 ml
Potassium bromide	90 g
Water	1 liter

Reversal Bleach

1. Mix together:

Water	1 liter
Potassium dichromate [K$_2$Cr$_2$O$_7$]	9.4 g
Sulfuric acid, conc.	12 ml

2. Bleach till clear

Mercuric Chloride Bleach for Reflection Holograms

1. Mix together:

Potassium bromide	25 mg
Mercuric chloride [HgCl$_2$]	25 mg
Water	1 liter

2. Bleach till clear
3. Rinse in water for one minute
4. Place bleached plate in fix until the plate turns from transparent to deep brown in color. At this point, the bleaching is completed.
5. Wash in water and dry.

CAUTIONARY NOTE: Mercuric chloride is a deadly poison.

Solution for Changing the Color of Reflection Holograms

Because of the position of the recorded interference fringes and the effect of drying (emulsion shrinkage), reflection hologram images can be seen with white-light sources (usually appearing to reflect green light). To change the color of the image from green to yellow to red, place the hologram in a solution of triethanolamine—ten parts water to one part triethanolamine—for about one or two minutes. Triethanolamine supposedly "fattens" the emulsion so that the recorded interference fringes expand in size to accommodate longer wavelengths of light. The color of the image can be controlled by the length of time the hologram is left to soak in the triethanolamine solution. A few minutes in the solution will convert the green color of the image to red.

APPENDIX II (C).   ALTERNATIVES TO SILVER-HALIDE EMULSIONS: DICHROMATED GELATIN FILMS

Although silver-halide photographic emulsions are the most common photosensitive materials used in making holograms, they are by no means the most efficient or ideal in this respect. In fact, so-called holographic plates and film made by Agfa-Gevaert and Kodak are simply high-resolution, fine-grain photographic emulsions developed not for holography but for other uses.

Silver-halide plates and film require time-consuming wet processing: developing, stopping, and fixing in noxious and poisonous chemical solutions. Besides, the use of silver-halide photographic emulsions necessitates great waste of chemicals (though silver can be periodically salvaged from the fix) since the solutions must be changed quite frequently. In fact, the principal reason silver-halide emulsions are used in making holograms is that the manufacturers have advertised to create a demand for their products. Silver-halide emulsions sensitized to red laser light have thus become popular because of their availability.

An alternative to silver-based holographic emulsions are the *dichromated gelatin* films and plates, usually sensitive to ultraviolet or blue light produced by an argon laser, but also sensitizable to red helium-neon laser light. Dichromated gelatin film is simply gelatin that is coated onto glass plates or acetate film, then sensitized to ultraviolet light by dipping or spraying on a solution of ammonium dichromate. Dichromated gelatin film is not available commercially but can be easily made in the usual manner of cookbook chemistry.

When a layer of gelatin coated onto a glass substrate and containing a small amount of dichromate is exposed to a laser-produced interference pattern of light, the illuminated portion of the plate becomes insoluble in water. This insolubility is caused by so-called *cross-linking* of gelatin molecules activated by the energy of the ultraviolet light. The cross-linking of gelatin is also called *hardening* or *tanning* of the gelatin. The gelatin that was not exposed to the interference pattern can be simply washed away in water. This also removes the unexposed dichromate. The film is then rapidly dehydrated by isopropanol, and the development is finished. Bleaching can be done by further exposure of the film to sunlight. If the dehydration of the gelatin is done too quickly, the gelatin tends to crack, resulting in "noisy" and dim holograms. When properly done, however, dichromated gelatin holograms are calculated to approach 100 per cent efficiency, which silver-halide emulsions can't even touch.

The following is a simple procedure for sensitizing glass plates or acetate film with dichromate. For more information about the mechanism of development of dichromates as well as a more comprehensive procedure, see Kosar's *Light Sensitive Systems* and "The Mechanism of Hologram Formation in Dichromated Gelatin" by R. K. Curran and T. A. Shankoff. *Applied Optics*. July 1970, Vol. 9, No. 7, p. 1651; and "Preparation of Dichromated Gelatin Films for Holography" by R. G. Brandes, E. E. Francois, and T. A. Shankoff. *Applied Optics*. November 1969, Vol. 8, No. 11, p. 2346.

MAKING DICHROMATED GELATIN FILM AND PLATES FOR HOLOGRAPHY

1. Mix 25 g USP gelatin (paper) in 500 ml tap water at room temperature (5% solution).
2. Heat the solution slowly (by the double-burner method) to 160 F then down to 140 F. Stir until clear with continuous stirring for about one hour.
3. Prepare 12.5 g ammonium dichromate in water to make a 4% sensitizing solution (12.5 × 25 = 312 ml water).
4. Take 1% sensitizing solution and add to the 500ml gelatin solution.

COATING GLASS PLATES WITH THE SENSITIZING SOLUTION

1. Using a syringe, spread 3 ml sensitizing solution onto a 4″ × 5″ plate. Drop solution in the middle of the plate and spread evenly.
2. Set the plates out to dry. The surface on which the plates are set must be level; check with a level to make sure, since unevenness will cause different thicknesses of gelatin in different parts of the plate.
3. Store plates by covering so dust and particulate matter do not settle on them.
4. Bake the plates to harden them (two to three hours at 275 F).

SENSITIZING THE PLATES

1. Using a red safelight, dip each baked plate into a 2% dichromate solution for two minutes. Take out, drain off, and lay flat. Be careful, since the emulsion is very soft.
2. Dip each plate into a half-and-half solution of 2% dichromate solution and isopropyl alcohol for one minute.
3. Dip each plate into a 90% solution of 2% dichromate solution to 10% of isopropyl alcohol for one minute.
4. Dip each plate into a solution of desiccated isopropyl alcohol for one minute.
5. Steam or air-dry each plate for 30 seconds.
6. The sensitized plates must be refrigerated and used within 24 hours of sensitizing.
7. The plates can be exposed to the laser-produced interference pattern, and a hologram will be formed.

DEVELOPMENT OF DICHROMATED GELATIN PLATES

1. Wash the exposed plate in a 10% methanol solution. Observe in dim white light.
2. Wash in desiccated isopropyl alcohol for one minute. Fix in potassium thiocyanate (30% solution) for five minutes.
3. Blow excess moisture off the plate with nitrogen gas (N_2).
4. Expose the plate to the reference beam only for about one minute.
5. Wash in the 2% solution of dichromate.
6. Wash in half-and-half solution of 2% dichromate and isopropyl.
7. Wash in 90% to 10% solution of isopropyl to 2% dichromate solution.
8. Wash in isopropyl alcohol.
9. Re-expose the plate to the reference beam.
10. Expose the plate to a sunlamp or sunlight after washing.

Sandbox Isolation Table And Holographic Studio

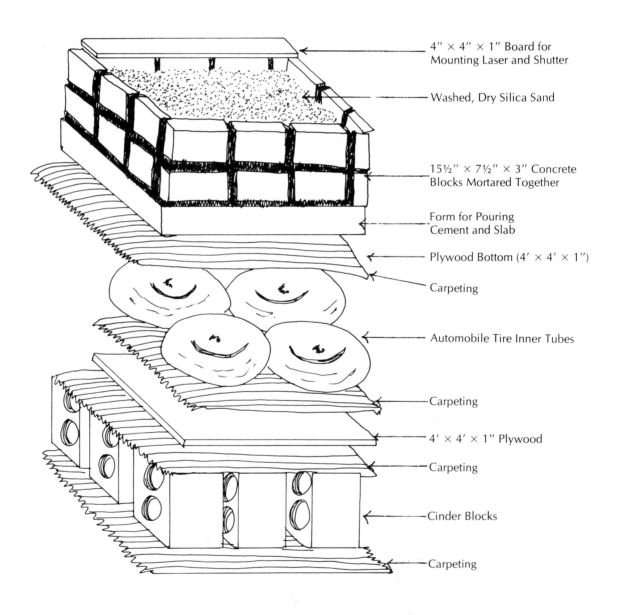

4″ × 4″ × 1″ Board for Mounting Laser and Shutter

Washed, Dry Silica Sand

15½″ × 7½″ × 3″ Concrete Blocks Mortared Together

Form for Pouring Cement and Slab

Plywood Bottom (4′ × 4′ × 1″)

Carpeting

Automobile Tire Inner Tubes

Carpeting

4′ × 4′ × 1″ Plywood

Carpeting

Cinder Blocks

Carpeting

Exploded-Perspective View After A Drawing By Peter Claudius

APPENDIX III (A).   HOLOGRAPHIC STUDIO CONSTRUCTION

When the interferometer test has been successfully completed, construction of the holographic studio can begin. The holographic studio should be located on a solid foundation, for example, a concrete ground floor in a garage or basement. Second-story locations and wooden floors are not recommended, since they are probably not stable enough to provide the vibration-free environment necessary for making holograms. At any rate, when the holographic studio is well isolated from people and traffic and the attendant noise, it is much easier to maintain the interferometric stability necessary for making holograms.

The entire room (the holographic studio) that houses the isolation table is in effect the holographic camera. Because of this, a lightproof enclosure must be built around the isolation table, or the entire room may be darkened. The location for the sandbox isolation table should be chosen carefully, possibly utilizing one or more of the walls of the existing room. Plenty of standing room on all sides of the table should be allowed for. To test the prospective location, the sandbox area can be drawn out in chalk on the floor prior to actually building it.

The sandbox isolation table consists of a cast-concrete slab with walls made of concrete bricks. This sandbox is then filled with fine-grade silica sand. To further shield this isolation table from vibrations, either several automobile tire inner tubes or one truck tire inner tube can be placed under the sandbox. The arrangement is commonly called an air-flotation system.

Classes at the School of Holography in San Francisco were initially taught on 4' × 4' sandbox-type isolation tables. This size has proven to be adequate although a larger-size table (4' × 8' or 5' × 10') allows larger holograms to be made as well as affording more area for experimentation with different holographic camera setups.

To build a 4' × 4' sandbox isolation table, first spread a layer of carpeting on the floor of the prospective sand-table area. The carpeting can be obtained free of charge in many cases from carpet stores as either remnants or samples. Set an array of cinder blocks up on end (three blocks on each side) on the layer of carpet. Spread another layer of carpet on top of the cinder blocks and cover it with a 4' × 4' sheet of 1-inch thick plywood. The plywood sheet is then topped by another layer of carpet on which four automobile tire inner tubes inflated ⅔ full are placed.

To make the concrete slab (the foundation for the air-supported sand table), construct a concrete mold by nailing a border of 2″ × 4‴'s around the outside edge of a 4' × 4' × 1″ piece of plywood. Attach another piece of carpet to the underside of the mold and put it on top of the inner tubes.

Lay out a lattice of 3-foot sections of iron reinforcement bar on the floor and wire the bars together where they cross. This structure will be cast into the concrete slab to give it additional strength.

A 4' × 4' × 4″ mold will hold about

4½ cubic feet of concrete. Mix the concrete according to the directions on the bag. A bag of Sacrete Concrete Mix that weighs 80 pounds will make an amount of concrete eight feet square by one inch thick. So, to make a slab 4' × 4' × 4" you need about eight bags, or about 640 pounds. Wet down the inside of the mold and spread a 1½ to 2-inch layer on the bottom. Place the lattice of the reinforcement bar in the middle of the box and add the rest of the concrete. Trowel the top smooth and allow the slab to dry.

Obtain twenty-four 15½" × 7½" × 3" concrete bricks and a sack of mortar from the local building supplier. Set a row of bricks around the edge of the slab to visualize how the bricks are to be placed. Mix the mortar a little at a time. Remove the bricks and lay a border of mortar 3 inches wide and ½-inch deep around the edges of one corner of the slab. Before applying the mortar to the bricks, moisten the edges of the slab with water and dip the bricks in water before laying them. Starting at the corner, mortar the bricks together around the edge of the slab until the first row is completed all around the slab. Then, add the second row of bricks and allow it to dry. When it is dry, mount a piece of 2" × 4" on the top of one side of the mortared brick wall by bolting it. Fill the sandbox with about 30 cubic feet of dry-washed silica sand (not wet sand). The sandbox isolation table is then ready for use.

APPENDIX III (B). CONSTRUCTION OF HOLOGRAPHIC STUDIO COMPONENTS

There are several different ways of mounting lenses, mirrors, and beam splitters in order to make sandbox holographic studio components. Probably the simplest method consists of using either black plastic pipe (ABS tubing or the kind used for plumbing) or aluminum tubing of 2-inch, 3-inch, or 4-inch diameter. Lenses and mirrors are then attached by gluing, bolting, or simply sticking the optics mounted on wood into the pipe.

Simple plate or film holders can be made from wood or particle board. The holographic plate then simply slides into place in the plate holder. Film holders can be made from Plexiglas, which acts as a support for the nonrigid holographic film.

Laser shutters can be made by attaching a slat of wood that can move back and forth in conjunction with a solenoid or a relay switch. When the laser shutter is in the "on" position, the slat of wood blocks the laser beam; when "off," the laser beam is allowed to pass. The laser shutter should be placed about six inches in front of the laser, where it can effectively block the laser beam.

Holographic Studio Components

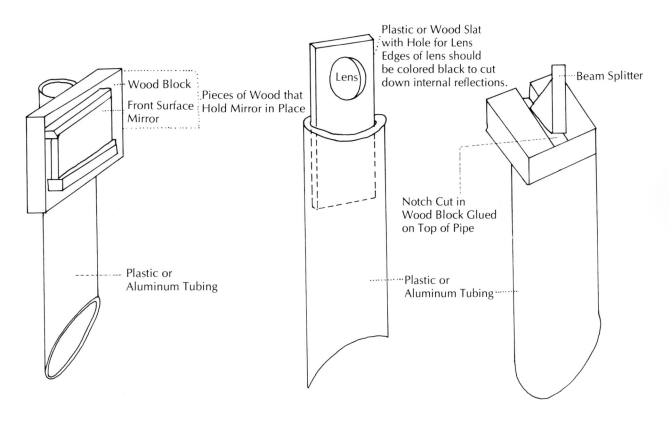

Wood Block

Front Surface Mirror

Pieces of Wood that Hold Mirror in Place

Plastic or Aluminum Tubing

Plastic or Wood Slat with Hole for Lens Edges of lens should be colored black to cut down internal reflections.

Lens

Plastic or Aluminum Tubing

Beam Splitter

Notch Cut in Wood Block Glued on Top of Pipe

Plastic or Aluminum Tubing

Holographic Plate Holder

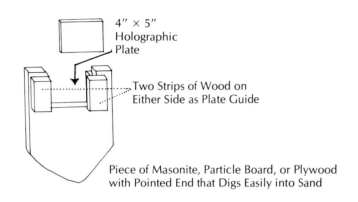

4″ × 5″
Holographic
Plate

Two Strips of Wood on
Either Side as Plate Guide

Piece of Masonite, Particle Board, or Plywood
with Pointed End that Digs Easily into Sand

Holographic Film Holder

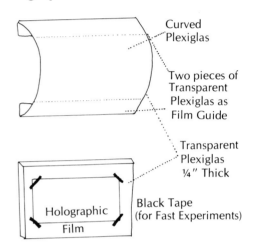

Curved
Plexiglas

Two pieces of
Transparent
Plexiglas as
Film Guide

Transparent
Plexiglas
¼″ Thick

Black Tape
(for Fast Experiments)

Holographic
Film

APPENDIX IV (A). ASSEMBLED LASERS

Assembled helium-neon and other gas lasers can be obtained from the following sources. For more information about available holographic supplies, check: *Laser Focus Buyer's Guide*, 385 Eliot St., Newton, Mass. 02164. Tel.: 617-244-2939.

Bausch and Lomb
Scientific Instruments Division
Bausch St.
Rochester, N.Y. 14602
  716-232-6000

Coherent Radiation
932 E. Meadow Dr.
Palo Alto, Ca. 94303
  415-328-1840

CW Radiation
111 Ortega Ave.
Mountain View, Ca. 94040
  415-969-9482

Hughes Electron Dynamics (Laser Products)
3100 W. Lomita Blvd.
Torrance, Ca. 90509
  213-534-2121 Ext 445

Jodon Engineering Associates
145 Enterprise Dr.
Ann Arbor, Mich. 48103
  313-761-4044

Metrologic Instruments, Inc.
143 Harding Ave.
Bellmawr, N.J. 08030
  609-933-0100

Optics Technology
901 California Ave.
Palo Alto, Ca. 94304
  415-327-6600

Spectra Physics Inc.
1250 W. Middlefield Rd.
Mountain View, Ca. 94040
  415-961-2550

APPENDIX IV (B). LENSES, MIRRORS, AND BEAM SPLITTERS

The following is a supplier of optics (lenses, front-surface mirrors, and beam splitters) used to make up the holographic components. Mirrors should always be tested with a laser for defects (such as so-called orange-peel aberrations) prior to buying them.

Edmund Scientific Co.
150 Edscorp Building
Barrington, N.J. 08007

APPENDIX IV (C). THE LIGHT METER

Light meters (photometers) of reasonable quality, though relatively expensive, can be obtained from:

Science and Mechanics
Instruments Division
229 Park Ave. South
New York, N.Y. 10003

A cheaper and easier obtainable version of a light meter is the Triplet V-O-M Meter with a cadmium-sulfide photocell that can be gotten from electronic supply stores.

To use the Triplet #310 meter as a light meter, the cadmium sulfide (CdS) photocell should be attached to the ohms scale of the meter. The attached photocell should be placed in a source of bright light, not necessarily a laser. The zero value of the scale should correspond with the maximum reading ×1000 on the ohms scale. When the cell is placed in the brightest light, the meter will indicate a low-resistance reading; this is exactly the opposite of the reading you get when you use a regular photometer. In other words, when the reading for the object beam is 60,000 ohms (60 Kohms) and the reading for the reference beam is 20,000 ohms (20 Kohms), the light ratio is 3 to 1, reference to object beam.

Using a Triplet meter, we can also derive an exposure chart, if we know the level of brightness and the relative light ratio. This means guessing, initially, using the scientific method of trial and error, until a pattern emerges that can be followed. It should be noted, though, that since we are using a makeshift light meter, we will produce a makeshift exposure chart. *There is no universally valid exposure chart for making holograms.*

APPENDIX IV (D). DARKROOM CHEMICALS

Most chemicals used in the darkroom for wet-processing holograms can be obtained in photographic supplies stores.

Developers:
Kodak D-19
Kodak Microdol-X
Acufine Developer

Stop Bath:
Kodak Indicator Stop Bath

Fixer:
Kodak Rapid Fixer

APPENDIX IV (E). HOLOGRAPHIC PLATES AND FILM

There are several brands of so-called holographic plates and film. Some of them were called high-resolution photographic plates for scientific purposes prior to the rise in interest in holography. The following are silver-halide-type holographic plates sensitive to 6328 Å red helium-neon laser light.

Agfa-Gevaert 8E75	holographic plates in 4″ × 5″ and 8″ × 10″	Kodak SO 173 film	See Kodak Publication J1: *Processing*
Agfa-Gevaert 10E75	sizes and film in 5″ × 150′ and 10″ × 150′ sizes (with or without antihalation backing).	Kodak #120 holographic plates	*Chemicals and Formulas* for black-and-white photography

Available from Agfa and Kodak at the following addresses. Write them for local dealers in your area.

Eastman Kodak Company
343 State Street
Rochester, N.Y. 14650

Agfa-Gevaert, Inc.
275 North Street
Teterboro, N.J. 07608

APPENDIX V. BIBLIOGRAPHY

HOLOGRAPHY:

Caufield, H. J., and Sun Lu. *The Applications of Holography.* New York: Wiley (Interscience), 1970.

Collier, R., C. Burckhardt, and L. Lin. *Optical Holography.* New York: Academic Press, 1971.

DeVelis, J. B., and G. O. Reynolds. *Theory and Applications of Holography.* Reading, Massachusetts: Addison-Wesley, 1967.

Kallard, T. (Ed.). *Holography: State of the Art Review.* New York: Optosonic Press, 1969, 1970.

Kock, W. E. *Lasers and Holography: An Introduction to Coherent Optics.* New York: Doubleday, 1969.

Schawlow, A. (Ed.). *Lasers and Light: Reprints from Scientific American.* San Francisco: W. H. Freeman, 1969.

Smith, H. M. *Principles of Holography.* New York: Wiley (Interscience), 1969.

Stroke, G. W. *An Introduction to Coherent Optics and Holography.* New York: Academic Press, 1969.

OPTICS:

Born, M., and E. Wolf. *Principles of Optics.* Oxford: Pergamon Press, 1970.

Fowles, G. R. *Introduction to Modern Optics.* New York: Holt, Rinehart and Winston, 1968.

Garbundy, M. *Optical Physics.* New York: Academic Press, 1965.

Jenkins, F. A., and H. E. White. *Fundamentals of Optics.* New York: McGraw Hill, 1957.

Strong, J. *Procedures in Experimental Physics.* Englewood Cliffs, New Jersey: Prentice-Hall, 1938.

LASERS:

Bloom, A. *Gas Lasers.* New York: Wiley.

Leinwoll, S. *Understanding Lasers and Masers.* New York: John F. Rider, 1965.

Lengyel, B. *Introduction to Laser Physics.* New York: Wiley, 1966.

Schawlow, A. (Ed.). *Lasers and Light: Reprints from Scientific American.* San Francisco: W. H. Freeman, 1969.

Steele, E. *Optical Lasers in Electronics.* New York: Wiley, 1968.

Van Pelt, W. F., *et al. Laser Fundamentals and Experiments.* Springfield, Virginia: Clearinghouse for Scientific and Technical Information, 1970.

BUILD YOUR OWN LASER:

Helium-Neon and Argon Lasers:
"The Amateur Scientist," *Scientific American,* September 1964, December 1965, February 1967, February 1969.

Dye Lasers:
"The Amateur Scientist," *Scientific American,* February 1970.

Pulsed Nitrogen Lasers:
"The Amateur Scientist," *Scientific American,* June 1974.

HOLOGRAPHIC RECORDING MATERIALS AND PROCESSES:

Katz, J., and S. Fogel. *Photographic Analysis.* Hastings-on-Hudson, New York: Morgan and Morgan.

Kosar, J. *Light-Sensitive Systems.* New York: Wiley, 1965.

Neblette, C. B. *Photography: Its Materials and Processes.* New York: Van Nostrand, 1962.

Thomas, W., Jr. (Ed.). *SPSE Handbook of Photographic Science and Engineering.* New York: Wiley (Interscience), 1973.

ELECTRONICS:

Ashe, J. *Electronics Self Taught With Experiments and Projects.* Blue Ridge Summit, Pennsylvania: Tab Books, 1971.

Basic Theory and Application of Transistors. TM 11-690, U.S. Government Printing Office, Washington, D.C., 1959.

Carson, R. S. *Principles of Applied Electronics.* New York: McGraw Hill, 1961.

Cathode Ray Tubes and Their Associated Circuits. TM 11-671, U.S. Government Printing Office, Washington, D.C., 1951.

Cleveland Institute of Electronics. *Electronics.* Cleveland, Ohio: C.I.E., 1964.

Lauer, H., R. N. Lesnick, and L. E. Matson. *Servomechanism Fundamentals.* New York: McGraw Hill, 1960.

Lurch, E. N. *Fundamentals of Electronics.* New York: Wiley, 1960.

Malmstadt, H. V., and C. G. Enke. *Electronics for Scientists.* New York: W. A. Benjamin, 1963.

Pettit, J. M. *Electronic Switching, Timing, and Pulse Circuits.* New York: McGraw Hill, 1959.

Richmond, A. E. *Calculus for Electronics.* New York: McGraw Hill, 1958.

Rojansky, V. *Electromagnetic Fields and Waves.* Englewood Cliffs, New Jersey: Prentice-Hall, 1971.

Theory and Use of Electronic Test Equipment. TM 11-664, U.S. Government Printing Office, Washington, D.C., 1952.

Training and Retraining, Inc. *Basic Electricity/Electronics.* Indianapolis, Indiana: Howard W. Sams, 1966.

Van Valkenburgh, Nooger, & Neville, Inc. *Basic Electronics.* New York: J. F. Rider, 1955.

———. *Basic Electricity.* New York: J. F. Rider, 1955.

Warschauer, D. M. *Semiconductors and Transistors.* New York: McGraw Hill, 1959.

GENERAL INFORMATION:

The Urantia Book. The Urantia Foundation, 533 Diversey Parkway, Chicago, Illinois, 1973.

Index